How Psychiatrists Make Decisions

International Perspectives in Philosophy and Psychiatry

Series editors: Bill (K.W.M.) Fulford, Lisa Bortolotti, Matthew Broome, Katherine Morris, John Z. Sadler, and Giovanni Stanghellini

VOLUMES IN THE SERIES:

Maladapting Minds: Philosophy, Psychiatry, and Evolutionary Theory
Adriaens and De Block

Portrait of the Psychiatrist as a Young Man: The Early Writing and Work of R.D. Laing, 1927–1960
Beveridge

Mind, Meaning, and Mental Disorder 2e
Bolton and Hill

What is Mental Disorder?
Bolton

Delusions and Other Irrational Beliefs
Bortolotti

Postpsychiatry
Bracken and Thomas

Philosophy, Psychoanalysis, and the A-Rational Mind
Brakel

Unconscious Knowing and Other Essays in Psycho-Philosophical Analysis
Brakel

Psychiatry as Cognitive Neuroscience
Broome and Bortolotti (eds.)

Free Will and Responsibility: A Guide for Practitioners
Callender

Reconceiving Schizophrenia
Chung, Fulford, and Graham (eds.)

Psychiatry Reborn: Biopsychosocial Psychiatry in Modern Medicine
Davies, Savulescu, Roache, and Loebel

Darwin and Psychiatry
De Block and Adriaens (eds.)

The Oxford Handbook of Philosophy and Psychiatry
Fulford, Davies, Gipps, Graham, Sadler, Stanghellini, and Thornton

Nature and Narrative: An Introduction to the New Philosophy of Psychiatry
Fulford, Morris, Sadler, and Stanghellini (eds.)

The Oxford Textbook of Philosophy and Psychiatry
Fulford, Thornton, and Graham

The Mind and its Discontents
Gillett

Psychiatric Neuroethics: Studies in Research and Practice
Glannon

The Abraham Dilemma
Graham

Is Evidence-based Psychiatry Ethical?
Gupta

Thinking Through Dementia
Hughes

Thomas Szasz: An Appraisal of His Legacy
Haldipur, Knoll, and Luft (eds.)

Dementia: Mind, Meaning, and the Person
Hughes, Louw, and Sabat (eds.)

Talking Cures and Placebo Effects
Jopling

Vagueness in Psychiatry
Keil, Keuck, and Hauswald

Philosophical Issues in Psychiatry II: Nosology
Kendler and Parnas (eds.)

Philosophical Issues in Psychiatry III: The Nature and Sources of Historical Change
Kendler and Parnas (eds.)

Philosophical Issues in Psychiatry IV: Classification of Psychiatric Illness
Kendler and Parnas (eds.)

Philosophy of Psychedelics
Letheby

Discursive Perspectives in Therapeutic Practice
Lock and Strong (ed.)

Schizophrenia and the Fate of the Self
Lopez-Silva and McClelland

Intruders in the Mind: Interdisciplinary Perspectives on Thought Insertion
Lysaker and Lysaker

Embodied Selves and Divided Minds
Maiese

Responsibility and Psychopathy
Malatesti and McMillan

Body-Subjects and Disordered Minds
Matthews

Rationality and Compulsion: Applying action theory to psychiatry
Nordenfelt

Diagnostic Dilemmas in Child and Adolescent Psychiatry
Perring and Wells (eds.)

Philosophical Perspectives on Technology and Psychiatry
Phillips (ed.)

The Metaphor of Mental Illness
Pickering

Mapping the Edges and the In-between
Potter

Trauma, Truth, and Reconciliation: Healing Damaged Relationships
Potter (ed.)

The Philosophy of Psychiatry: A Companion
Radden

The Virtuous Psychiatrist
Radden and Sadler

Addiction and Weakness of Will
Radoilska

Autonomy and Mental Disorder
Radoilska (ed.)

Madness and the Demand for Recognition
Rashed

Feelings of Being: Phenomenology, psychiatry and the sense of reality
Ratcliffe

Experiences of Depression: A study in phenomenology
Ratcliffe

Recovery of People with Mental Illness: Philosophical and Related Perspectives
Rudnick (ed.)

Values and Psychiatric Diagnosis
Sadler

The Oxford Handbook of Psychiatric Ethics
Sadler, Van Staden, and Fulford

Madness and Modernism: Insanity in the light of modern art, literature, and thought
Sass

Disembodied Spirits and Deanimated Bodies: The Psychopathology of Common Sense
Stanghellini

Lost in Dialogue: Anthropology, Psychopathology, and Care
Stanghellini

One Century of Karl Jaspers Psychopathology
Stanghellini and Fuchs

Emotions and Personhood
Stanghellini and Rosfort

Essential Philosophy of Psychiatry
Thornton

The Oxford Handbook of Psychotherapy Ethics
Trachsel, (eds)

Naturalism, Hermeneutics, and Mental Disorder
Varga

The Healing Virtues: Character Ethics in Psychotherapy
Waring

Empirical Ethics in Psychiatry
Widdershoven, McMillan, Hope and Van der Scheer (eds.)

The Sublime Object of Psychiatry: Schizophrenia in Clinical and Cultural Theory
Woods

Alternate Perspectives on Psychiatric Validation: DSM, ICD, RDoC, and Beyond
Zachar, St. Stoyanov, Aragona, and Jablensky (eds.)

Intruders in the Mind: Interdisciplinary Perspectives on Thought Insertion
Lopez-Silva and McClelland

Vice and Psychiatric Diagnosis
Sadler

Philosophical Perspectives on Psychedelic Psychiatry
Letheby, Gerrans

Anatomy of an Avatar: The virtual self in pathologies of self awareness
Philip Gerrans

How Psychiatrists Make Decisions

The Science of Clinical Reasoning

Douglas W. Heinrichs

OXFORD
UNIVERSITY PRESS

OXFORD
UNIVERSITY PRESS

Great Clarendon Street, Oxford, OX2 6DP,United Kingdom

Oxford University Press is a department of the University of Oxford.
It furthers the University's objective of excellence in research, scholarship,
and education by publishing worldwide. Oxford is a registered trade mark of
Oxford University Press in the UK and in certain other countries

Published in the United States of America by Oxford University Press
198 Madison Avenue, New York, NY 10016, United States of America

British Library Cataloguing in Publication Data
Data available

Library of Congress Control Number: 2024953147

ISBN 9780198960706

DOI: 10.1093/9780198960720.001.0001

Printed and bound by
CPI Group (UK) Ltd, Croydon, CR0 4YY

Foreword

Skilled clinical reasoning and problem-solving are central to both virtuous practitioners as well as educators and trainees. Psychiatric educators are in the business of training skilled, conscientious practitioners by supervising the trainees' active clinical work with patients. Psychiatric education includes didactic education in lectures and seminars as well. Assessments of trainees' progress vary from programme to programme and country to country, but often include observations of clinical encounters and feedback about clinical skills and knowledge. However, the thinking processes involved over the arc of the treatment relationship are overlooked; assessments of clinical skills are cross-sectional, but every treatment course is longitudinal over the course of many clinical encounters. Even in the special case of psychotherapy supervision, discussion of clinical problem-solving is framed typically within the parameters of the therapy 'orientation' in the terms particular to the psychotherapy theory. The synthetic work is not addressed: that of assimilating medical, psychopharmacological, social context, patient preferences and values, and the many psychotherapies into nuanced, multidimensional treatment plans responsive to the evolving circumstances and viewpoints of the patient. Psychiatric trainees are left to themselves to figure out this crucial skill set on their own. Douglas Heinrichs' splendid book addresses this problem through assembling a scientifically credible approach to a multimodal, longitudinal clinical reasoning for the doctor–patient relationship.

I provide a bit of background. Approaches to clinical reasoning, despite a recent but modest bump in scholarly activity and publications, remain undertheorized, undiscussed, and worst of all, untaught to trainees (Sadler and Hulgus 1992). Perhaps the most commonly discussed and most popular approach to clinical reasoning today is evidence-based medicine (EBM) (Sackett *et al.* 2000). EBM begins prior to the patient encounter. The clinician, in effect, is expected to be familiar with the scientific literature pertinent to a particular diagnostic category, synthesize it, and critically appraise it in accordance with a hierarchical strength-of-evidence rubric touted by the EBM authors. By the time the EBM clinician encounters the patient with said diagnosis, she then applies the previously assimilated diagnostic and treatment knowledge to the unique patient. EBM acknowledges that 'patient values' may enter into the treatment planning and negotiation, but there the

EBM discussion of values ends. Moreover, with EBM, the clinical-reasoning guidance largely ends after this initial diagnostic appraisal. The evolution of treatment involving the navigation of idiosyncratic side effects, drug interactions, health literacy, medication hesitancy and non-adherence, psychosocial stressors, and most other clinical problems, get ignored in EBM. Yet experienced clinicians know that addressing these issues has as much, or more, to do with successful outcomes than selecting the most strongly supported 'evidence-based treatment'.

In reaction to EBM is the values-based practice approach articulated and elaborated by Fulford and colleagues (Handa and Fulford 2023). With VBP, the issue of patient and clinician values is addressed directly and linked as a complementary partner with scientific evidence and findings. Articulating the concept of 'dissensus', VBP aims to aid clinical decision-making through providing a clinical space where values-related agreement and disagreement can be addressed and clinical action taken, even without 'consensus' or resolution of conflict. VBP then represents a cluster of practice skills and humanistic principles important to, but not the same as, clinical reasoning.

In psychiatry, treatments of clinical reasoning have addressed the broader context of care more directly. David Brendel (2006), drawing from American pragmatism, formulates a '4p' model of clinical problem-solving, providing a scaffold for decision-making in the face of clinical problems. Clinicians should consider the Practical dimensions of all scientific inquiry, the Pluristic nature of clinical phenomena, the Participatory elements involving multiple people relevant to care, and the Provisional nature of scientific explanations. Nassir Ghaemi's (2003) work on methodological pluralism in clinical problem-solving emphasizes matching the right clinical 'tool' for the right clinical problem. Through multiple examples he shows how this matching can be done. Bradley Lewis addresses clinical problem-solving through his Narrative Psychiatry (2011) approach, which elevates the role of the patient in clinical reasoning. Like his predecessors McHugh and Slavney (1998), Lewis frames clinical problems through multiple 'perspectives', though for Lewis these perspectives are formed as multiple stories that doctors and patients can use to understand the clinical problem(s). These stories, whether biomedical or personal–historical, once shared in the treatment relationship, can be reformed in ongoing dialogue with the patient.

With this book, Douglas Heinrichs assimilates these formulations while transcending them. Importantly, Heinrichs has presented an approach to clinical problem-solving that draws from empirical evidence-bases, narrative unfolding, decision analysis, and doctor–patient relationship-building in providing a tailored treatment formulation for each unique patient—which, in turn, is revised recursively in the longitudinal treatment relationship. I think the centre of Heinrichs' approach is his casting the individual clinician as a scientist, formulating a testable theory about each patient that is matched to the unique characteristics the patient brings to the relationship. The 'ingredients' for the 'theory' of each patient is the evidence provided by the patient's history-giving, experiences, and behaviours, while the clinician brings the pluralistic knowledge and experience-base in the various sciences of psychiatry. Jointly with each patient, Heinrichs develops the treatment relationship as a unique single-case study design aimed at understanding and then intervening with the patient's clinical problems. Here, the doctor–patient dialogue, laboratory studies, and examination findings allow him to test hypotheses about his theory of the patient, adding refinement and nuance as the treatment relationship unfolds over time. The single-case study is the process of working collaboratively, the patient providing a validating element as the theory is revised recursively over the treatment arc. The patient and doctor are peers in the constitution of a healing theory of the patient. In this process, the patient's values are not just an extraneous ingredient requiring special consideration, but a seamlessly integrated component in formulating a credible theory of the patient's clinical problems. The clinician and patient, then, in this context of tracking the forces driving clinical problems, can re-make the problems towards aiding the patient.

Heinrichs' approach requires a mature and well-trained clinician to realize its maximum potential. If one does not have good training, knowledge, and experience in psychopharmacology and the various psychotherapies, then one cannot engage the patient with these new interpretations of their clinical problems, much less recognize how they converge and sediment into dysfunctional patterns. In this sense Heinrichs' approach also supports, perhaps requires, a clinician's commitment to 'life-long learning', because as the clinician's knowledge and experience grows, the more versatile the clinician becomes in formulating new hypotheses to test in the clinical encounter with the patient as an 'epistemic partner'. Heinrichs' clinical-reasoning approach is growth-promoting for clinicians as well as patients. One might say Heinrichs provides a path to clinical wisdom.

This book is a thrilling contribution to the clinical-reasoning literature, and I'm excited to see how its applications and impact will grow as it is read and appreciated widely in training programmes, clinic settings, and beyond.

John Z. Sadler, M. D. Director,
UT Southwestern Program in Ethics in Science & Medicine
Professor of Psychiatry

The Daniel W. Foster, M.D. Professor of Medical Ethics
Distinguished Teaching Professor
University of Texas Southwestern
Dallas, Texas

References

Brendel, D. H. 2006. *Healing Psychiatry: Bridging the Science/Humanism Divide.* Cambridge, MA: M.I.T. Press.

Ghaemi, S. N. 2003. *The Concepts of Psychiatry: A Pluralistic Approach to the Mind and Mental Illness.* Baltimore: The Johns Hopkins University Press.

Handa, A., & B. K. Fulford. 2023. Values-based practice: a theory-practice dynamic for navigating values and difference in health care. *Royal Institute of Philosophy Supplements* 94, 219244.

Lewis, B. 2011. *Narrative Psychiatry: How Stories Can Shape Clinical Practice.* Baltimore: The Johns Hopkins University Press.

McHugh, P. R., & P. R. Slavney. 1998. *The Perspectives of Psychiatry, second edition.* Baltimore: JHU Press.

Sackett, D. L., S. E. Straus, W. S. Richardson, W. Rosenberg, & R. B. Haynes. 2000. *Evidence-Based Medicine: How to Practice and Teach EBM, second edition.* Edinburgh/New York: Church-Livingstone.

Sadler, J. Z. & Y. F. Hulgus. 1992. Clinical problem solving and the biopsychosocial model. *American Journal of Psychiatry* 149(10):1315–1323.

Preface

This book grew out of reflections on my experience of over 40 years as a practising psychiatrist, first as a full-time clinical researcher and teacher in an academic setting, and then in the private practice of general psychiatry, in which I treated a wide range of patients in both inpatient and outpatient settings. I accrued this experience during a period of rapid and fundamental changes in the way psychiatry understands itself and how it articulates standards of practice. I will say much more about this process in Chapter 1. Here let me just say that it has left me with several deep-seated concerns.

The paradigmatic activity of the psychiatrist is the treatment of the individual patient who seeks relief from suffering or dysfunction. Central to this work is making rational and effective decisions about treatment interventions based on the psychiatrist's understanding of the patient's situation in all its unique complexity. Yet for all its importance, this decision-making process has not been well characterized. While it is informed by more generalized knowledge drawn from formal research or accumulated clinical wisdom, we often see how we arrive at an understanding of a unique individual as the intuitive product of our experience, the 'art' of medicine.

My undergraduate concentration was in philosophy. Among its most fundamental set of questions are those comprising epistemology: what are the methods by which we acquire valid knowledge about the world and how justifiable is our confidence in the resulting beliefs based upon those methods? This book is an attempt to answer those questions with respect to clinical reasoning and decision-making in psychiatry. I argue that this is of considerably more than just theoretical interest. In my time in psychiatry, I have seen a growing determination in the field to see itself as scientific, whatever that is taken to mean, and a valuing of the results of formal clinical and basic research over any unique understanding the clinician has of the individual patient seeking help. The rise to prominence of evidence-based medicine (EBM) and practice with its articulated methodology threatens to replace the traditional approaches to clinical reasoning as the guiding standard as to how we should practice. To salvage any credibility for traditional modes of clinical reasoning, we need to articulate its underlying methodology and carefully examine the basis for our confidence in the alleged knowledge about our patients that comes from such reasoning and that guides our decision-making process in treatment. I will argue that without such a methodology, EBM by itself fails to provide an adequate framework for translating research findings into good clinical care. I will further argue that articulating a methodology of clinical reasoning can have further benefits for enhancing the training of future psychiatrists and the enhancement of expertise in all mental health practitioners.

The primary audiences for this book are psychiatric residents and those responsible for the education of future psychiatrists. I believe it also can be of considerable value to psychiatrists at all stages of their careers who are interested in enhancing their expertise. It is equally relevant to other mental health clinicians. I also hope it has value for those philosophers with an interest in the conceptual underpinnings of psychiatry. The philosophy of psychiatry has in fact become a vigorously growing subspecialty in applied philosophy. Yet in my experience this vibrant discipline is largely unknown to most psychiatrists.

It is easy to think about philosophy as an abstract and self-referential intellectual activity with little relevance to real-world concerns. But at its most basic, philosophy is an attempt to rigorously examine the largely tacit conceptual assumptions that underlie any complex human activity. A field like psychiatry is especially rich in these, given that we deal with some of the most complex issues about human nature. Are our natures best construed in terms of our biology or as something irreducibly mental? To what degree should we be held responsible for our own actions and to what degree are we the products of forces—be they biological, psychological, or social—that are beyond our control? What is mental illness and how is it best characterized? How do we reconcile our third person knowledge of our patients with their first-person experiences? The list goes on. We cannot help having underlying conceptual assumptions about these issues that guide our concrete behaviour as we do our work. If we choose not to wrestle with them explicitly, that is to do philosophy, we will act on them implicitly. And we will often find that our implicit concepts are to varying degrees erroneous and incomplete. This can result in seriously misguided decision-making and behaviour in our clinical work.

This book would never have happened without the wisdom and support of many individuals over many years. The teachers and supervisors in my residency at the University of Cincinnati taught me the importance of listening carefully to my patients in trying to understand their unique circumstances. My time as both a fellow and faculty member at the Maryland Psychiatric Research Center, and especially the generous mentorship of William T. Carpenter, Jr., provided me with a solid grounding in methods of clinical research. Since leaving academic medicine for private practice, I have found my scholarly home among the members of the Association for the Advancement of Philosophy and Psychiatry. This group of philosophically oriented clinicians and philosophers with a particular interest in the field of psychiatry has contributed greatly to the development of the ideas in this book. I have presented drafts of various parts of this work to the group, where I received both generous encouragement and rigorous constructive critique. Both were quite needed. And, of course, this book would not exist without the privilege of working with my patients over these many years. I have been repeatedly struck by their willingness to persevere in the face of great suffering and difficulties and to engage in the challenging collaboration needed to improve their lot.

In addition, there are several individuals I want to acknowledge for their specific role in this project. Scott Waterman and Robyn Bluhm read early versions of this manuscript and provided many useful insights and suggestions, as well as engaging in extensive informal discussions over the years that this work was percolating. My daughter, Victoria Marshall, provided badly needed technical support. I want to express my very special appreciation to John Z. Sadler. He encouraged me to persevere with this work over many years, and then most generously gave a tremendous amount of time and energy to rigorous and critical readings of the manuscript both in its early stages and again near its completion. The book is vastly better because of his help. The limitations are solely my own. Finally, I could not have done this project without the immense love, patience, and support of my wife, Diane, and my family.

Contents

SOCRATES: Well, here is a question for you. Suppose someone came up to your friend Eryximachus or his father Acumenus, and said, 'I know how to treat the body in ways which allow me to raise or lower temperatures, to get people to vomit, to make their bowels move, and so on and so forth—whatever I choose or decide is best. And since I have this knowledge I regard myself as a professional doctor, and I claim to be able to make others doctors too by imparting this knowledge to them.' How do you think they would respond to this speech?

PHAEDRUS: I am sure they would ask him whether he also knew whom he should treat in these ways, and when, and how much.

SOCRATES: And what of he said, 'No, I do not. But I claim that anyone who learnt these treatments from me would be able to do what you ask'?

PHAEDRUS: I think they would say that the man was out of his mind, and was imagining that he had become a doctor after having heard someone reading from some book or other, or having accidentally come across some minor drugs, but that he really had no understanding of this area of expertise.

Plato, Phaedrus

Epigraph reproduced with permission from Waterfield, R. (trans.). 2009. *Plato: Phaedrus*. Oxford, UK: Oxford University Press.

Chapter 1
Introduction

The need to articulate a method of clinical reasoning in psychiatry

1.1 Clinical reasoning as decision-making

This book articulates and characterizes a process of clinical reasoning in psychiatry. Clinical reasoning here refers to the cognitive processes employed by a psychiatrist in making treatment decisions regarding a specific patient. This includes both general decisions about the sort of treatment modalities to employ, as well as more specific decisions such as exactly what medication to prescribe or psychotherapeutic intervention to make. A good psychiatrist needs many other skills, such as how to best gather clinically relevant information and how to establish and maintain a good working alliance with patients. This book focuses specifically on the act of clinical decision-making.

How does clinical decision-making articulate with the full range of competencies employed by the practising psychiatrist? Although there is some overlap, we can conceptualize clinical training in psychiatry as focused on acquiring four categories of knowledge and skill. First, the psychiatrist must become familiar with a broad body of general information in a range of disciplines that inform the practice of psychiatry. This includes general medicine, psychology, neuroscience, biology, epidemiology, cognitive science, sociology, and anthropology. Especially relevant are psychopathology and therapeutics. Some of this knowledge comes from familiarity with the research literature and textbooks, but especially in the case of psychopathology and therapeutics, also from the collective experience of practitioners in the field. Much of this is conveyed in training by supervisors and clinical instructors. This sort of knowledge is typically expressed as generalizations about groups of patients or a certain sort of patient. Over time, psychiatrists gain enough experience to develop some of their own generalizations that apply to certain types of patients and clinical situations.

How Psychiatrists Make Decisions. Douglas W. Heinrichs, Oxford University Press.
DOI: 10.1093/9780198960720.003.0001

Second, the psychiatrist learns a range of skills and techniques for eliciting specific and detailed information from an individual patient. These include clinical interviewing skills, history taking, conducting a mental status examination, and establishing a working therapeutic alliance in which the patient feels comfortable sharing information about themselves. In contrast to the knowledge derived from the study of psychopathology, therapeutics, and other informing disciplines, this activity yields a great deal of highly specific and idiosyncratic information about an individual patient. This skill set also includes the ability to organize the vast array of information into conceptually meaningful units that allow the clinician to develop a manageable understanding of patients and their current situation.

The third skill set is using the knowledge from the first two, both generalized and specific to that patient, to formulate an informed understanding of the individual patient's distress in a manner that allows the psychiatrist to posit and select clinical interventions to ameliorate the problems. Clinical reasoning as discussed in this book focuses on this third activity, i.e. making clinical decisions. However, it is important throughout to remember that the foundational knowledge derived from the first two sets of skills is absolutely necessary for making good clinical decisions. Without them, clinical decision-making would become an arbitrary and reckless activity.

The fourth skill set is the ability to effectively implement treatment interventions once they have been selected. While obviously an important component of a psychiatrist's competencies, we will give it very little attention in this book. Our focus will be on the rational process of selecting what interventions to make.

Clinical reasoning in medicine has been practised for centuries. So why is it important to reflect on it now? Historically clinical reasoning and decision-making has been something practised, often tacitly, with very little effort to explicate its nature or the cognitive method behind it. New practitioners have generally learned it by modelling from their teachers combined with an array of anecdotes, rules of thumb, and clinical pearls; whereas the actual methodology of clinical reasoning has been relatively opaque and poorly understood. That was never a good thing but it has become an increasingly serious problem in our current environment of psychiatric practice and training. Now we have some urgency to articulate the process of clinical reasoning in psychiatry in an explicit way, to allow its rational basis to be understood and justified, and to provide a way to distinguish good from bad clinical reasoning.

1.2 Shifting psychiatric paradigms

We can appreciate the nature of this urgency by looking at the changes in psychiatry since I entered it as a young resident in the mid-1970s. At that time the field was in a state of major transition away from a conception of psychiatry dominated by psychodynamic thinking, with its emphasis on psychotherapy as the privileged treatment modality and psychological theorizing as the primary explanatory framework, to perspectives built on exciting new developments in psychopharmacology and neurochemistry that focused on understanding psychiatric phenomena using the methodologies of the biological sciences.[1] Departments of psychiatry varied on how rapidly they shifted and on whether they attempted to maintain some balance between these approaches or to become strong advocates of one or the other. My own residency experience was split between two programmes, the first of which was a strongly psychoanalytic department, followed by a programme that included both approaches but with a much stronger biological orientation.

Our use of language can be misleading here and a disclaimer is in order. By contrasting 'psychological' with 'biological', I do not mean to imply that I am taking a stand on whether the phenomena of concern to psychiatrists are mental things or physical things—what philosophers would call a question of ontology. I personally think that this is a false distinction based on a mind-body dualism dating back at least to Rene Descartes in the seventeenth century.[2] Rather I am referring to a methodological distinction; i.e. whether a particular psychiatric phenomenon is most productively explored with the methods of biological sciences or psychological ones. Throughout this book, unless indicated otherwise, I will use terms like biological, neurological, organic, psychological, and environmental in this methodological sense without ontological assumptions.

A marked shift in the way that psychiatric teaching was done and the relative status of various activities in most departments of psychiatry accompanied this transition. Previously many prominent department chairpersons and senior faculty were known as excellent clinical educators. Their reputations often came from their high quality of instruction of residents and their ability to develop an understanding of an individual case, often demonstrated in grand rounds settings. Following this transition, most are known for their research activities. Status is related to the ability to get research grants and to publish research studies. The clinical education of residents often falls increasingly to junior faculty. There also have been changes in our practices of clinical thinking. Psychodynamic work pivotally involved the case formulation. This was an effort to develop a highly individualized understanding of each patient considering a wide range of relevant factors, albeit with a disproportionately strong emphasis on psychodynamic issues.

What we now see in its place is evidence-based medicine (EBM).[3] Here the primary task of clinicians is to determine how to best apply results from research studies, especially controlled clinical trials, conducted on groups of patients judged to be similar to the individual they are trying to treat. Although lip service is paid to the overall complexity of treating the individual patient and the need for clinical judgement, in truth the application of results from research studies and clinical trials is treated as the pivotal task. Other aspects of the clinician's work may be necessary but are seldom discussed with much depth or with any articulation of the processes and methods involved. Rather these are dismissed as the 'art' of medicine rather than the science. Clearly the science is more valued. Clinical reasoning about the individual patient is often viewed as a stopgap process to be employed where knowledge from clinical studies is not yet available.

1.3 The devaluing of clinical judgement

As I look at how psychiatry has evolved over my career, I am struck by two forms of growing pessimism frequently voiced by clinicians, The first is a devaluing of what is the clinician's primary activity as clinician. Too often today, any exercise of clinical judgement that is not a direct inference from a controlled clinical trial is given little credibility. Recently, I experienced a particularly prosaic instance of this when a patient was denied coverage by his private insurer (I practice in the United States) for a more expensive antidepressant even though he had shown a dramatic and sustained relief from a severe and chronic suicidal depression after starting the drug, having failed several other antidepressant trials. The official explanation given was that there were cheaper antidepressants he had not yet tried and that in clinical trials the response rates were as good as with the more expensive drug. Since the likelihood of a response was the same, he should be tried on the cheaper drug. In my appeal to the insurance company, I pointed out the obvious fact that for this particular patient the odds of response were not the same, noting that the patient's odds of responding to the cheaper drug was no more than 60–70% based on the clinical trials and actually lower given that he had failed trials of several similar drugs.[4] By contrast, his odds of responding to the more expensive drug were 100%, since he was already on it and had shown a dramatic and sustained benefit. I eventually won the appeal, but the blindness to even so obvious a piece of information based on the individual experience of the patient in deference to data from clinical trials is disturbing.

In short, attempts to formulate an understanding of the specific individual patient beyond the application of clinical trial data is acknowledged as necessary but is seen as having clearly inferior value and credibility. This is embodied in the 'hierarchy of evidence' that is so central to the practice of EBM.[5] This is a tricky issue in that lip service is always paid to the need for the clinician to consider the full range of factors affecting the individual patient. But in practice this is viewed as unscientific, as an art. The implication is that eventually when we know enough this form of stopgap reasoning will become obsolete, or at least its role will be severely curtailed. This issue relates strongly to psychiatry's sense of inferiority as a science. Psychiatry has always been viewed as more impressionistic and less based on hard science than the rest of medicine. One could argue that much of the appeal of biological psychiatry is that it makes the psychiatrist feel scientific.[6]

However, what is science? What is the boundary that distinguishes scientific method from other, presumably inferior, forms of thinking? It is a common assumption that there cannot be a science of the individual patient, since science is thought to involve inferences drawn from large numbers of cases studied in a rigorous manner and subjected to statistical analyses. One of the arguments of this book will be that limiting science to this sort of aggregated data is based on a very flawed notion of what constitutes valid scientific method. The underlying assumption is that science is a system for discovering laws of nature that are ideally universal or at least applicable across some specified domain by a method of inference involving large numbers, followed by applying deductive reasoning from these generalizations to the individual case.[7] For many contemporary philosophers of science this view of science is problematic and far too narrow.[8] Since at least the 1970s a new model of science has evolved that argues that general laws play a much less important role in scientific thinking than the development of models that attempt to map on certain concrete pieces of the real world. Fleshing out this notion will take some work and will be argued in more detail in Chapter 10. We will see that the result is quite compatible with a method of clinical reasoning in psychiatry that involves generating models for individual patients. I argue that what the clinician does is not outside of science but is in fact an instance of its paradigmatic activity.

1.4 Disappointing progress in psychiatric knowledge

The impact of this first form of pessimism is magnified by a second one. For if one looks at the results of what is normally viewed as science in psychiatry—results that form the evidence base that clinicians are supposed to use to make clinical decisions about their patients—it is hard not to be deeply disappointed. In clinical trials the magnitudes of effect sizes and the portion of patients responding tend to be disappointingly modest.[9] It is also disconcerting how similar the responses are between a wide range of drugs and psychotherapies across groups, but with extreme variability of responses between individuals. The noise swamps the signal. Differential efficacy is difficult to demonstrate for almost anything in psychiatry. Furthermore, the clinical trials are based on patients grouped by diagnostic categories that themselves are highly arbitrary and questionable. Since the DSM III (*Diagnostic and Statistical Manual III*), diagnoses have become increasingly reliable; that is, different clinicians can apply the same criteria and come to the same diagnostic assignment of a given patient. Do those categories have any valid meaning? Do they constitute real groupings in the world? Do they cut nature at its joints? The growing consensus is that the answer to these questions is 'no'.[10] If the groupings themselves are not valid, how much faith can be put in studies based upon them?

Of course, the field has the perennial hope that increased understanding at some basic level of neuroscience will provide a solid key for psychiatric science. Each new candidate is greeted with waves of enthusiasm followed by gradual disillusionment.[11] Initially it was the chemistry of monoamines. Later it was receptor physiology. The neurochemistry of inflammation is thought to offer promise. Perhaps genetics offers great hope. Most recently the mapping of neural circuitry has received tremendous attention. I am not denying that some useful information and applications have accrued from these areas of study. Yet none of them have in fact yielded the tremendous breakthrough that was originally promised. The belief that some single level of analysis will hold the key to understanding psychiatric disorders and developing effective treatments may simply be naïve given the tremendous complexity and interactive qualities of the nervous system and the person. Perhaps we should not be surprised that the vast bulk of the major breakthroughs in psychiatric therapeutics up to now has been the result of serendipity as opposed to flowing from any scientific understanding.

As work in all these areas proceeds, the field has become a tower of Babel. Isolated results continue to pour in from many sources, but how they relate to each other and to the clinical care of patients is unclear. Bits of work in what has been called translational research represent laudable efforts at some sort of piecemeal integration, but this work is in its infancy and how productive it will be is yet to be determined.[12] A recent initiative by the National Institutes of Mental Health, the Research Domain Criteria (RDoc), is an attempt to structure a meaningful integration of research work at various levels.[13] Yet in spite of its protestations to the contrary, the disproportionate emphasis on neural circuitry as the key level of analysis has led many critics to be sceptical and to see RDoc as just another wolf of reductionist thinking in the sheep's clothing of integration.[14] In any case the verdict is still out on this approach.

Clearly the contemporary psychiatrist, charged with being scientific in the application of current knowledge to the care of the individual patient, operates on a sea of shifting sand. All the decades of work trying to elucidate basic mechanisms of psychopathology in terms of neurotransmitters, receptors, genetics, or neural circuitry have to date provided remarkably little concrete guidance in alleviating the suffering of the patients we treat. The body of work that constitutes controlled clinical trials of various treatment modalities, mostly pharmaceuticals, is also remarkably unhelpful. Many results fail to replicate.[15] Any given treatment rarely shows consistent differential efficacy compared with other treatments. Over time many treatments prove less efficacious than the results from initial studies suggested. Psychiatrists often must choose from among treatments that have demonstrated modest efficacy at best but significant potential side effects. Increasingly, the funding for treatment trials comes from the pharmaceutical industry. This raises not only concerns about the potential bias of existing studies, but the financial interests of the industry dictate what gets studied and what does not. Furthermore, most studies have quite poor external validity. This refers to the confidence one can have in applying results from the study to other patients who were not in that study. When one looks at the exclusion criteria for most research trials, most of the patients seen by a psychiatrist would not qualify to participate in the very studies that the psychiatrist is asked to draw upon in deciding how to treat that patient. Typically, then, psychiatrists are thrown back on their own clinical judgement and reasoning. However, they have been told that this process is impressionistic, unscientific, and at the very bottom of the 'hierarchy of evidence' in its value or trustworthiness.[16] Is it any wonder that pessimism abounds?

1.5 A role for philosophy

What follows is my effort to confront this pessimism. This effort is the result of reflecting on my 40 years as a psychiatrist, starting with my training in the mid-1970s during the great transition in psychiatry, through a decade in academic medicine as a full-time clinical researcher and teacher, followed by 40 years of clinical practice in both inpatient and outpatient settings. Before entering medicine, my undergraduate work concentrated in philosophy, and my abiding interest in the philosophical underpinnings of psychiatry has greatly influenced my views as developed in this book.

Philosophy is all too often viewed as an esoteric and abstract activity with little relevance to the practical doings of life.[17] However, at its most basic, philosophy is simply a thoughtful attempt to articulate and critically examine the often-unspoken premises and conceptual foundations upon which we base our view of the world or some portion of it. To choose not to consciously philosophize does not mean that we do not have premises and foundations to our thinking. It just means that they are unconscious and uncritically accepted.[18] If they are wrong or limited, as they often are, they cause considerable mischief in our practical dealing with the world. The comments in section 1.4 about the implications of terms like 'biological' or 'psychological' is an example. The philosopher Mary Midgley likens philosophy to plumbing.[19] Its subject matter lies under the floorboards or behind the walls of our cognitive constructs. We do not think about it until something goes wrong—water leaks through or something starts to stink. Then we must dig down to the problem, find the malfunction, and fix it. Philosophy is about finding the conceptual malfunctions when some aspect of our activity is not working well and otherwise inexplicable problems appear.

1.6 Two audiences

This book is aimed at two audiences. It primarily addresses psychiatrists, psychiatric trainees, and psychiatric educators as well as other mental health clinicians. It has concrete and practical applications to how we as psychiatrists conduct our task, how we judge its adequacy, and how we train new generations of psychiatrists to continue the work. To understand the rationale for this approach some excursions into philosophical topics are necessary. I urge the psychiatric reader to be patient with these, as their relevance to the clinical goals will become clear. I will keep technical philosophical jargon to a minimum and no prior philosophical training will be presumed. These topics will be considered only to the extent necessary to make the larger argument.

The second audience is philosophers with an interest in the conceptual foundations of the discipline of psychiatry. I hope that this book will provide some useful ideas as to how their conceptual work in the growing field of the philosophy of psychiatry relates to the concrete challenges facing psychiatric clinicians in their day-to-day work. To avoid excessive disruption to the flow of the discussion, I will at times use endnotes to provide background information that may be needed by one set of readers but be well-known to the other.

1.7 A roadmap

Before starting on this journey, a brief roadmap of where we are going may be helpful. As the aim of this book is to articulate a methodology of clinical reasoning and decision-making in psychiatry, we will first consider why such an effort is necessary at all. Because EBM offers itself as a method to guide a clinician's decision-making process, and is currently the leading candidate for such a role, Chapter 2 asks whether EBM and evidence-based practice in fact provide an adequate model of how clinicians should reason about treating individual patients. I contend that EBM in practice falls far short of providing such a basis for clinical reasoning. Given the prominence of this approach in current thinking in psychiatry, and medicine in general, the matter will be explored in considerable detail to show the specific and serious shortcomings of this approach. I do not deny that EBM has value in its original goal of helping the busy clinician navigate the vast and always growing clinical literature. It clearly does. The overreach of offering itself as a comprehensive method for making clinical decisions is where it falls short.

In Chapters 3 through 6, each chapter is organized around a case study of an individual treated by the author. These phenomenological explorations are used to articulate an explicit methodology of clinical reasoning and decision-making that characterizes the actual if tacit process employed by good clinicians to develop an individualized understanding and treatment strategy for each patient. The goal is to make that tacit process explicit.[20] The contrast with the contribution of EBM approaches in these cases will be highlighted.

In Chapter 7 I pull together the insights from these case studies to generate a general picture of clinical reasoning and decision-making in psychiatry. The process of clinical reasoning involves the construction of a working model of the individual patient. Such models are derived from those factors that contribute to the problems that brings the patient into treatment and organized around the patient's goals. These models can be characterized as the elaboration of **patterns of propensities** or dispositions and their pathways of potential interactions; hence, I call them **patterns of propensities (POP) models**. These propensities and pathways can suggest various points of therapeutic intervention for the psychiatrist that are tailored to the individual patient. This first function of POP models is descriptive, that is making explicit the reasoning process that experienced clinicians usually do automatically and implicitly. It is also normative in that this method is recommended as an effective and credible strategy for understanding individual patients and formulating treatment decisions.

I will then make the case for a second, pedagogical function for POP models. By making the process of clinical reasoning explicit clinicians can better assess the adequacy of their own efforts and improve the quality of their clinical work. Having such an explicit understanding of clinical methodology can also greatly aid in the teaching of new clinicians entering the field. Starting with the assumption that the goal of psychiatric education is to further the development of expertise, in Chapter 8 we will summarize our current best understanding of human decision-making in general and what we can know about how experts in a range of fields make decisions in situations of time pressure and incomplete and ambiguous information. This field of research has yielded an understanding of expert reasoning that is strikingly like the POP model approach. This same body of research has suggested strategies that can be employed to enhance the development of expertise.

In Chapter 9 this understanding of human decision-making and expertise will be applied specifically to psychiatry. Some general notions will be put forward that could provide the basis for a programme of pedagogical advances in psychiatric education for both neophytes in training and for established clinicians committed to the ongoing enhancement of their expertise.

In Chapter 10 we will consider a third function for the POP model. Can POP modelling serve as a basis for confidence in the conclusions of clinical reasoning about an individual patient as providing valid knowledge? This function is one of epistemological justification. Science and the application of scientific method are viewed as the strongest support for the credibility of proposed knowledge in many areas of contemporary life. Medicine in general and psychiatry in particular are no exceptions. Indeed, much of the devaluing of clinical reasoning about the individual patient can be tied to the belief that this activity cannot be scientific. But this assumption is rooted in a very specific and limited view of what constitutes valid science that dominated thinking about science in the twentieth century until the 1960s. From that vantage, explaining an individual case scientifically involves seeing it as an instance under some generalized 'law' that is itself derived by inductive inference from many instances.[21] This view, derived from the philosophical school of logical empiricism, still dominates much of the thinking in medicine and psychiatry, but has largely been replaced or modified by current thinkers in the philosophy of science. This more recent understanding will be utilized as a basis for the argument that a clinical methodology based on the construction of models in fact constitutes an instance of the paradigmatic activity of science as it is actually practised across all fields, even in such 'hard' sciences as physics.[22] From this perspective it will be possible to distinguish the science of psychiatry proper from closely related and informing disciplines with which it is often confused, in particular, psychopathology and clinical therapeutics.

Chapter 11 will summarize the argument of this book and emphasize its relevance to the clinical practice of psychiatry proper. I hope that this explicit method of clinical reasoning and decision-making in psychiatry will help clinicians practice more creatively and rigorously, with greater confidence in the validity of our knowledge, and with enhanced outcomes for our patients.

Endnotes

1. Luhrmann 2001
2. Descartes 1641
3. Sackett *et al.* 2000; Straus *et al.* 2019
4. Agency for Healthcare Research and Quality, 2010
5. Montori and Guyatt 2008; Guyatt *et al.* 2001, p. 7
6. Guze 1989
7. Hempel 1965
8. Suppe 1989; Cartwright 1983, 1999; Giere 1988, 1999; van Fraassen 1980, 2008
9. Swartz *et al.* 2008; Warden *et al.* 2007
10. Mayes and Horwitz 2005; Cooper 2004; Kirk *et al.* 2015; Poland and von Eckardt 2013; Cuthbert and Insel 2013; Kinghorn 2020; Ghaemi 2003
11. Morrissey and Goldman 1984
12. Fort *et al.* 2017; Rubio *et al.* 2010
13. Cuthbert and Insel 2013; Casey *et al.* 2013
14. Lilienfeld 2014; Frances 2014
15. Miczek and Robbins 2019
16. Phillips and Sadler 2021
17. Wallace *et al.* 1997
18. Aftab and Waterman 2021
19. Midgely 2005, pp. 146–147
20. Polanyi 2009
21. Hempel 1965
22. Suppe 1989; Cartwright 1983, 1999; Giere 1988, 1999; van Fraassen 1980, 2008

Chapter 2
Does evidence-based medicine provide a clinical method?

2.1 Assessing evidenced-based medicine as an adequate clinical method

My intention in this book is to argue for a specific methodology for clinical reasoning and decision-making in psychiatry. Some of the reasons as to why this is an important task have been elaborated in Chapter 1. Before beginning on this journey, we might question the need for a fresh understanding of clinical method at all. In recent years the notion of evidence-based medicine (EBM), and its elaboration as evidence-based practice, has gained considerable prominence by promising a foundational methodology for the clinician to practice state-of-the-art individualized treatment.[1] In fact, its dominance in this area justifies an extensive critical examination of how well it lives up to its promise before arguing that an alternative framework is necessary. Here I provide a selective critique of EBM focusing specifically on its ability to provide a methodology for clinical reasoning and decision-making for the psychiatrist treating an individual patient. A more extensive critique of evidence-based medicine applied to psychiatry can be found in *Is Evidence-based Psychiatry Ethical?* by the psychiatrist and ethicist Mona Gupta.[2]

Philosophers of science Robyn Bluhm and Kristin Borgerson,[3] in a thoughtful review of the history and development of EBM, trace the trajectory of its growing ambition. Initially EBM was proposed as a method to help the clinician efficiently access relevant clinical studies from the overwhelming volume of published studies and reports. It is hard to argue with the usefulness of this effort. Every clinician knows that it is literally impossible to keep up with all the developments in the field, particularly if one is engaged in a busy clinical practice. Textbooks and other printed material rapidly become out of date. New information is pouring in at a rate that is truly unmanageable.

How Psychiatrists Make Decisions. Douglas W. Heinrichs, Oxford University Press. © Oxford University Press (2025).
DOI: 10.1093/9780198960720.003.0002

Over time evidence-based medicine sought to do more than provide a framework for rational information retrieval and assessment and offered itself as a comprehensive practice framework to guide the clinician's reasoning and decision-making in the treatment of individual patients. In terms of the framework used in section 1.1, EBM has moved from focusing on the first skill set of acquiring generalized knowledge, to claiming that it provides an optimal methodology for the third skill set, making clinical decisions about an individual patient. With this step the question of the adequacy of evidence-based practice arises.[4] Bluhm and Borgerson point out that part of the growing success and popularity of EBM comes from the rhetorical value of the name itself.[5] After all, in an era where psychiatry, along with the rest of medicine, seeks scientific legitimacy and empirical support for its treatments, how can one argue against something that defines itself as evidence-based? Certainly no one would want to argue for clinical judgements and treatments that are not based on a consideration of evidence.

Furthermore, when one looks at the professed mission of EBM, one is hard-pressed not to concur with the intent. Unless otherwise noted, all quotations from the EBM literature will be from a highly respected and classic representative text by Sackett *et al.*, *Evidence-Based Medicine: How to Practice and Teach EBM*[6] or the updated fifth edition of that text by Straus *et al.*[7]

> Evidence-based medicine (EBM) is the integration of best research evidence with clinical expertise and patient values. By *best research evidence* we mean clinically relevant research, often from the basic sciences of medicine, but especially from patient-centred clinical research into the accuracy and precision of diagnostic tests (including the clinical examination), the power of prognostic markers, and the efficacy and safety of the therapeutic, rehabilitative, and preventative regimens . . . By *clinical expertise* we mean the ability to use our clinical skills and past experience to rapidly identify each patient's unique health state and diagnosis, their individual risks and benefits of potential interventions, and their personal values and expectations . . . By *patient values* we mean the unique preferences, concerns and expectations each patient brings to a clinical encounter and which must be integrated into clinical decisions if they are to serve the patient.[8]

At first blush, this mission statement would seem to accord due respect to the individualized needs of the patient. However, the question is whether this is substantive or rhetorical. We must look at the actual implementation guidelines of EBM to know.

2.2 The hierarchy of evidence

As a first consideration, what exactly constitutes valid and relevant evidence in treating the individual patient? EBM holds that while evidence comes in many forms, a clear hierarchy exists in the quality of evidence that should guide the clinician in making decisions.[9] In the realm of treatment decisions, the crown in this hierarchy is the double-blinded, randomized, controlled, trial (RCT).[10] Below this are other kinds of naturalistic or observational studies.[11] Beyond that are case report series. The clinical experience of the individual practising clinician is at the very bottom of this list. The vast gulf between the value placed on RCTs compared to every other kind of evidence is reflected in the advice on efficient searching for evidence from the literature. 'If the study was not randomized, we would suggest that you stop reading it and go on to the next article in your search . . . Only seek out the conclusions of observational studies if you cannot find any randomized trials'.[12] Thus the presence of one RCT would preclude the consideration of even a vast number of observational studies and case reports that might contradict that RCT. At first glance all of this may appear quite reasonable and self-evident. After all, the point of randomized clinical trials is to control for certain kinds of potential biases and to isolate important variables. How could that not be the most important form of data and lead to the best treatment decisions? But RCTs carry their own biases and limitations.

2.3 Biases in what gets subjected to clinical trials

The treatment that has the strongest evidence base from existing RCTs may have that strong base for reasons other than being the best treatment. Many relevant clinical questions could be tested by RCTs yielding useful information. However, conducting such trials is very expensive and time-consuming. Given the limits imposed by time and money, most relevant clinical questions will never be subjected to randomized clinical trials. Ironically, one of the important clinical questions that has never been subjected to rigorous testing with RCTs is the efficacy of the evidence-based medicine approach itself to improving patient outcomes. This strikes me as a serious omission by their own criteria that the advocates of EBM dismiss rather cavalierly.[13]

So, what factors drive the questions addressed by clinical trials? First and foremost are likely to be economic considerations. Clinical trials that hold the promise of generating substantial financial benefit to the entity conducting the trial are likely to be pursued. This undoubtedly explains the high proportion of clinical trials conducted by pharmaceutical companies on new drugs or on drugs still under patent protection.[14]

In addition to economic factors, the ease of standardization of the relevant variables under study often determine what gets studied. Treatments that can be highly standardized are much easier to study than treatments that are not. In the case of psychotherapy trials for instance, modalities such as cognitive-behavioural therapy (CBT) lend themselves to structured manuals of treatment, making it easier to assure that the same treatment approaches are applied uniformly, if not necessarily well, to all subjects. Psychotherapies derived from psychoanalytic and psychodynamic approaches, on the other hand, are much more difficult to standardize and are less likely to be studied. Thus, in the case of psychotherapy trials, the fact that CBT has the best support from RCTs in no way can be interpreted to indicate that it is the best treatment for all. The state of the evidence more likely reflects that it is easier to standardize and study this treatment and hence to demonstrate benefit in clinical trials over other psychotherapeutic modalities.

The ease of standardization also applies to the conditions being treated in treatment trials. Thus, many more clinical trials address the treatment of bipolar I disorder than bipolar II disorder.[15] This is true despite bipolar II disorder being considerably more common in the population.[16] The decisive factor is likely that bipolar I disorder is easier to diagnose consistently and reliably, whereas bipolar II disorder fades off into other affective disorders in a more spectrum-like manner.[17]

2.4 Variation in individual responses and wilful ignorance

In psychiatric clinical trials, the difference in aggregate response rates between various treatments are typically relatively modest and a sizeable portion of individuals fail to respond. In this situation the question which EBM seeks to answer—all things being equal which treatment is most likely to succeed with a particular type of patient?—may not be the most crucial question. After all, for any given patient all things in fact are not equal. With modest differences in response rates between treatments and sizeable non-response rates to all treatments, these other factors which differentiate individual patients may often contribute more to predicting response than the main effect of the treatment modalities under study. While randomization and sufficient statistical power guard against individual differences systematically skewing the main effect at the group level, a crucial question is whether factors known to the clinician make a particular patient, Ms Jones for instance, more likely to respond to one treatment or another. In essence there are two different questions here that can easily be conflated and treated as if they are the same question. The first question is: Given a group of patients sharing a defined set of characteristics, usually a common diagnosis and a set of exclusion criteria, what treatment is likely to be beneficial to the highest percentage of this group? The second question is: What treatment intervention is likely to be optimal for this individual, Ms Jones, who happens to have this diagnosis and meets this set of exclusion criteria? These two questions are only synonymous if we know absolutely nothing else about Ms Jones except that she has this diagnosis and meets the set of exclusion criteria. If that is all we know then the answer to the first question would be our best guess as to the answer to the second question.

It is a well-known but underappreciated fact that the methodology of clinical trials and the presumptions of the statistical methods used to analyse them require the investigator to actively ignore a wide array of individualized facts about participating subjects that would be readily available to the clinician treating them. In fairness, advances in statistical methods, made possible by the power of computers to perform complex data-heavy calculations, increasingly allow investigators to go beyond main effects and identify ever more complex interactive effects. Meaningful subgroups of patients can thereby be identified that may have differential treatment responses.[18] Nonetheless, only variables characterizing subgroups with a fairly large number of members are likely to be identified by these techniques. Rather than telling us the likelihood of Ms Jones responding to a given treatment, clinical trials give us an idealized test of an intervention for a purified group of patients, usually in quite unnatural conditions. Its value is that if the treatment fails in these idealized conditions, we have little reason to expect benefit in less-than-ideal conditions.[19]

A critical and insightful appraisal of this aspect of statistical analysis in research trials and how it evolved in the history of statistics can be found in a work by the statistician Herbert Weisberg appropriately entitled *Willful Ignorance: The Mismeasure of Uncertainty*.[20] Weisberg stresses that our current techniques for quantifying probability evolved in the seventeenth century from mathematical explorations of games of chance, such as dice. In these highly artificial and contrived circumstances, every effort is made, assuming the game is fair, to assure that there is nothing knowable about any specific role of the dice except the general rules of the game. The background assumptions of the statistical methods employed assume this pervasive ignorance. As probability techniques began to be applied to other real-world problems, those assumptions of pervasive ignorance about specific individual instances remained. However, in the real world all instances of a category of events are not designed to be interchangeable and generated by the same rules. Life is not a game. Weisberg's argument is that it is often informationally wasteful and misleading to ignore the extensive specific information we have in real-world situations just so we can apply a statistical methodology that was never really designed for this kind of circumstance. He furthermore argues, and I think quite correctly, that this choice to ignore information about the individual that is available to the clinician has, in medical research, the effect on clinicians of questioning the relevance of research studies to real-life practice and of feeling in turn that their knowledge and expertise is devalued and trivialized by the research community.

In real-life clinical situations, it is almost never the case we know nothing else of relevance about Ms Jones. We are likely to know many individual features of Ms Jones that are plausibly relevant to her response to one or another treatment intervention. Admittedly, most of this additional knowledge is highly unique to this individual patient and hence is not directly supportable by information from studies, including RCTs. However, does this mean that this individualized information has no evidentiary value? Weisberg goes on to argue that with the rise of mathematical probability and its gradual application beyond games of chance to real-world situations, there has been a gradual change in and narrowing of the concept of probability itself. The older concept of probability involved a judgement of the likelihood of an event based upon a consideration of all relevant factors for and against the likelihood of that event. Most of these factors could not be easily or meaningfully assigned a precise numerical value. Nonetheless it was assumed that a rational person could make an assessment with reasonable confidence. One setting in which this sort of probability is still explicitly employed is in our legal system. Here we ask juries or judges to use all available information to assess probability of culpability in terms such as the preponderance of the evidence, probable cause, or beyond a reasonable doubt, without an appeal to mathematical probability statistics.

2.5 Existential bias and external validity

We are now ready to look beyond the facile acknowledgement of the need to consider individual factors in the mission statement of EBM and ask precisely how one is expected to proceed in implementing this. When one considers the primary texts for the application of EBM, instructions on how to proceed appear to imply that this kind of individualized information is of so little value that it can be largely ignored.

The highly structured procedure for applying EBM is practice has five steps, the first of which is to ask an 'answerable question'.[21] Based on the examples provided, these questions, in the area of choice of treatment, typically take the form of the first question in section 2.2.2.3; namely, what is the best treatment, based on available evidence, for patients suffering from condition X. (Sometimes evidence is available to allow the question to take the form of an interaction; i.e. what is the best treatment for patients with condition X who also have attributes Y and Z. Nonetheless, the question is still framed in terms of a group of patients, albeit a somewhat narrower one.) Once one reviews the evidence and answers the group-level question, in general one proceeds to recommend that treatment with a specific patient suffering from condition X. The few exceptions have to do with obvious issues such as medical contraindications or history of allergic response to the treatment in question. The more complicated question about the patient's preferences and values will be considered in a later step.

However, note that virtually no consideration is provided for the possibility that the clinician may know individualized facts about a given patient that would influence the probability of the response to the proposed treatment. Yet given the high number of patients who fail to respond to virtually all treatments in psychiatry,[22] these individualized factors are likely to have a tremendous impact on outcome, probably greater than the magnitude of difference in group response between the various treatments being considered. Any attempts to systematically study in significant numbers even the more commonly shared of such factors would require colossal sample sizes, and colossal budgets, that are simply unavailable. Furthermore, many are part of patterns of interacting factors unique to a specific patient and hence not addressable at a group level of analysis, no matter how fine-grained, what Sadler has termed 'existential bias'.[23] Thus, the clinician must decide how to weigh these factors apart from reliance on formal research studies.

This issue of idiosyncratic factors dovetails with the larger problem of external validity in applying clinical trial data to individual patients. The concept of external validity refers to the question of how valid it is to assume that the results obtained from a group of patients in a clinical trial will apply to an individual not in that trial. Clearly the strongest case for external validity would occur in a situation where the individual would have in fact met the criteria for participation in the study being applied. At the grossest level one can ask the question whether the patient has the same diagnosis as the group of people in the clinical trial. This seemingly simple question is immensely complicated by the arbitrary and fuzzy nature of diagnostic categories in psychiatry. Most clinical trials are designed to test a treatment for a specific diagnosis. Hence the strength of EBM is heavily dependent upon the adequacy of our diagnostic system. This poses a huge set of problems. There has been an ongoing and extensive critique of the current diagnostic system in psychiatry, DSM (*Diagnostic and Statistical Manual*), currently in the fifth edition.[24] Since the emergence of a paradigm shift with DSM-III in 1980,[25] the emphasis in diagnosis has been on grouping patients based on clusters of symptoms without a clear unifying concept of a disease entity. This reflects in part the tremendous lack of knowledge of underlying psychopathological processes in psychiatry as well as the competing theoretical systems within the psychiatric community. The official DSM position is that we cannot at this time decide upon these underlying questions given our current level of ignorance, and hence the best strategy is to have an atheoretical, purely empirical approach to grouping patients. The resulting system however usually consists of a cluster of symptoms, some number of which a given patient must have to meet the diagnostic criteria. However multiple patients can meet the diagnostic criteria with clusters of symptoms that are quite unlike each other. This raises the real question of whether they have the same condition. As a result of all this it may be far from self-evident that Ms Jones, who meets criteria for condition X, sufficiently resembles the patients in a given clinical trial of condition X to have confidence that the results of that trial would apply to her.

The underlying commitment of recent editions of the DSM to a totally atheoretical approach in fact appears to be contradicted in the DSM itself. For a high percentage of diagnoses one of the criteria for applying the diagnosis in question is that the symptoms cannot be 'better explained by another condition'. This is a curious criterion indeed. The clinician is called upon to consider an alternative explanatory mechanism for the symptoms when the presumption of the diagnostic system itself is that there is no basis for any explanations at all. One cannot choose between alternative explanatory hypotheses without positing some underlying mechanism for the condition. Furthermore, that mechanism would have to be based upon some theoretical notion about what causes the condition. So, for instance, with the patient who is manifesting high anxiety and paranoid ideation, is the anxiety explainable in terms of the paranoia which is then posited as the more fundamental underlying process? Alternatively, are high levels of overall anxiety posited as the underlying process that can lead to some paranoid ideation as a secondary phenomenon? There is no atheoretical way to decide between these two options yet the DSM is committed to being atheoretical while at the same time demanding this judgement of the clinician in applying the DSM criteria.

The problems with external validity however go even further. Virtually all studies have in addition to the inclusion criteria for diagnosis, a series of exclusion criteria. These criteria serve to exclude research participants who have the targeted diagnosis, yet are considered inappropriate for the study. Exclusion criteria can vary immensely. They frequently exclude patients who have other psychiatric comorbidities or perhaps other medical comorbidities. They may exclude patients with substance abuse issues or patients who are deemed too dangerous to be entered into the clinical trial due to suicidality or violence. There may be age restrictions or gender restrictions. The usual rationale for these exclusion criteria is to generate a reasonably purified or idealized sample representing condition X, uncontaminated by the messy complications of the real world. In most cases the clinician is left with the unsettling dilemma of attempting to use clinical trial data to make decisions about treating a specific patient who likely would have been excluded from those very clinical trials. How reasonable is it to apply the results of such a trial to Ms Jones? It is very striking that in one of the standard textbooks on applying EBM the statement is made that in general these exclusion criteria, and the fact that the patient being treated would not have qualified for the study, is generally not important and the concern can be dismissed.[26] Ironically this grand dismissal is made without offering any evidentiary support whatsoever. Furthermore, the assumption that this matter can be disregarded is highly counterintuitive. After all the exclusion criteria are there for a reason, so presumably they matter in some important way or at least were thought to matter by the people conducting the trial. This cavalier dismissal may reflect the fact that a concession on this point would deeply weaken the rationale for the presumed efficacy of the EBM agenda as a guide to treating individual patients.

2.6 The lure of pseudo-quantification

Another set of issues relating to external validity surrounds the severity of the condition in question. Conditions in psychiatry, like the rest of medicine, typically occur with varying levels of severity of the same condition in different patients. Often the level of severity matters a great deal in determining the most appropriate and effective treatment for a given patient. The shifting severity of individuals participating in clinical trials over time is also likely to have a significant impact on the outcome of the studies. Hence the increasing placebo response rates in more recent medication trials in psychiatry may reflect the inclusion of higher numbers of less severely ill individuals in these trials. As a result, generating a separation between active treatment and placebo has become increasingly difficult.[27] So an important factor related to external validity involves whether the severity of the condition of Ms Jones is comparable to the severity in the group of patients in the clinical trial being considered. Supporters of EBM offer a ready answer to this question in the form of standardized rating scales, arguing that administering such scales both to subjects in a clinical trial and Ms Jones herself would allow a clear and precise determination of comparative severity.

At first glance this seems like a very sensible answer to the question. However, a question arises that threatens to undercut the value of much of our clinical trial data in psychiatry: Do rating scales in general represent meaningful quantification of illness severity or rather constitute pseudo-quantifications. Applying numbers in describing anything confers tremendous psychological power. It immediately suggests high levels of precision, accuracy, and objectivity, and offers the potential for the application of data analytic strategies that are thought to enhance our understanding of the phenomenon in question. However, whether this promise is fulfilled in any given case depends upon the validity of the quantification process itself. Do the numbers measure in any meaningful way the phenomena in question? In many aspects of medicine this is a simple issue to assess. When we record a patient's blood pressure, we know precisely what we are measuring. There may be room for differing interpretations of the significance of those measurements, but the meaning of what we are measuring is clear. If a systolic blood pressure is 120 mmHg, we know precisely what that means. We also know that a systolic blood pressure of 80 mmHg is one-third less than a reading of 120 mmHg. The same is true in measuring concentrations of various substances in a blood sample, in measuring cardiac output with an echocardiogram and in many other medical tests.

However, what does a number on a particular symptom rating scale in a psychiatric trial mean? It is frankly difficult to know. We can demonstrate that different raters have been trained to use a particular instrument with good reliability, that is that they tend to agree in assigning a numerical rating of severity of a given symptom in a given patient. However, what is the validity of that rating; if the patient moves from, for example, a rating of six to a rating of four on a particular rating scale, has the symptom severity dropped by a third? If a patient moves from a rating of four to rating of two is that the same magnitude of change as moving from a six to a four? The truth is we have in general not the foggiest idea. Yet these numerical data points are subjected to statistical analysis and form the basis of decisions about treatment efficacy as if they were clear objective measurements like blood pressure or sodium concentration.

Beyond this are the whole set of problems as to which measurements reflect the severity of the condition in question most meaningfully. If one is employing a depression rating scale with 10 items, do all those items carry equal weight and importance in determining the overall severity of depression? If a given treatment tends to reduce items 4, 7, and 9 and another treatment tends to reduce items 2, 8, and 13 by the same magnitude, which one is more important in treating the condition of depression and for which patients? Furthermore, rating scales over time may evolve to measure precisely those dimensions that existing treatments are known to affect the most. A long-standing concern has been that one of the most popular depression rating scales, the Hamilton Rating Scale for Depression,[28] excessively emphasizes symptoms known to respond to existing antidepressant medication while the Beck's Depression Inventory[29] focuses on cognitive aspects of depression that are addressed by his CBT, both in ways that are disproportionate to their importance in the total depressive syndrome.[30]

2.7 Applying EBM to the individual patient: Step by step

Given the cumulative impact of the aforementioned limitations of EBM, how useful is this approach for decision-making by a clinician treating a specific patient. In short, is an alternative clinical decision-making methodology needed? Before offering such a method, let us walk through the steps proposed by EBM for decision-making with the individual patient. What is described as the 'full-blown practice of EBM' comprises five steps.[31] The first step involves 'converting the need for information (about prevention, diagnosis, prognosis, therapy, causation, etc.) into an answerable question'. The second step is 'tracking down the best evidence with which to answer that question'. The third step relates to 'critically appraising that evidence for its validity (closeness to the truth), impact (size of the effect), and applicability (usefulness in our clinical practice)'. Step four is 'integrating the critical appraisal with our clinical expertise and with our patient's unique biology, values and circumstances'. Finally, step five involves 'evaluating our effectiveness and efficiency in executing steps 1–4 and seeking ways to improve them both for the next time'. A few observations are in order regarding each of the steps.

2.7.1 Step 1: Formulating answerable questions

The first step, formulating answerable questions, has already been discussed in section 2.2.2.3. As noted, the natural and relevant question of the practising clinician, namely how to best proceed in evaluating and treating this specific patient, Ms Jones, must be transformed into a question about some group of patients within which Ms Jones could be appropriately included. How well do the answers to these 'answerable' questions provide optimal guidance in answering the clinician's fundamental question about Ms Jones? As we have seen, there are multiple reasons to be doubtful in this regard. Beyond this, the challenge is how to work from the general nature of an answerable question back to an individualized understanding of an individual patient. EBM promises to address this matter in the fourth step.

2.7.2 Steps 2 and 3: Finding and appraising the relevant evidence

The second step relates to strategies for tracking down best evidence for answering the questions posed. This of course was the original mission of EBM and it certainly addresses a critical need for the busy clinician confronted with the mass of ever-growing information. As originally conceptualized step two is closely related to step three, which emphasizes the critical need for the clinician to be able to evaluate the methodological adequacy of any trials relevant to one's question. EBM certainly serves to highlight the generally poor preparation clinicians receive in being able to critically assess study design and implementation, including statistical analyses. One of the underlying premises of EBM is that the clinician should not take judgements of experts on faith but ask how good the evidence is. As EBM has gained popularity, this step has introduced a considerable irony regarding the role of experts. A whole series of summary resources have appeared both in print but largely online in which a group of 'experts' search the relevant literature and provide succinct summaries of studies including assessment of the strengths and weaknesses of the studies.[32] The spokespersons for EBM strongly encourage the use of these sources as an extremely time efficient way for the busy clinician to get relevant information. Now the clinician, instead of oneself assessing the quality of information, is once again trusting some group of experts to make those judgements. Ironically, the third step is essentially delegated.

> 'Preappraised evidence resources for clinical decisions are built according to an explicit process that values research according to both its scientific merit ('hierarchy of evidence') and its readiness for use in clinical decisions . . . Only if we can't find what we want there would we need to tackle the harder task of searching larger bibliographical databases, such as MEDLINE, and applying the critical skills that are taught in this book'.[33]

2.7.3 Step 4: Integrating evidence with the patient's unique biology, values, and circumstances

The fourth step is the particularly crucial one for our concerns. This step promises to guide us in using our clinical expertise and our knowledge of the patient's unique biology, values, and circumstances to decide how applicable and relevant results from research studies are in treating Ms Jones. The way step 1 defines an answerable question moves us away from our individual patient to a consideration of groups of patients. If EBM is to function as a framework and comprehensive clinical methodology for the practitioner, step four must lead us back to that individual patient. The guidance as to how to do this as provided in step four must be explicit, comprehensive, and reasonable. EBM tackles this issue by asking four questions.

2.7.3.1 Can study results be applied to this patient?

The first question is whether our patient is so different from those in the study that its results cannot apply to her. Given the importance of this issue, the answer of EBM is quoted at some length:.

> One approach would be to demand that our patient meet all the inclusion criteria for the study and to reject its usefulness if our patient didn't meet each and every one of them. This is not a very sensible approach because most differences between our patients and those in trials tend to be quantitative (they have different ages or social classes or different degrees of risk of the outcome event or of responsiveness to therapy) rather than qualitative (total absence of responsiveness or no risk of the event). We'd suggest that a far more appropriate approach is to consider whether our patient's socio-demographic features or pathobiology are so different from those in the study that its results are useless to us and our patient; only then discard its results and resume our search for relevant evidence. There are only a few occasions when this might be the case: different pharmacogenetics, absent

> immune responses, comorbid conditions that prohibit the treatment, and the like. As a consequence of this clinical (as opposed to actuarial) approach, it's rare that we have to toss away a study for this reason . . . Sometimes treatments appear to produce qualitative differences in the response subgroups of patients so that they appear to benefit some subgroups but not others. Such qualitative differences in response are extremely rare . . . Unless the difference in response makes biological sense, was hypothesized before the trial, and has been confirmed in a second, independent trial, we'd suggest that you accept the treatment's overall efficacy as the best starting point for estimating its efficacy in your individual patient.[34]

Several aspects of this argument are problematic. What is the evidence that individuals excluded from a clinical trial would respond similarly to those included in the trial? The casual dismissal of the concern that our patient, Ms Jones, might not meet all the inclusion criteria for the study under consideration seems unwarranted and lacks face validity. While some inclusion criteria can relate to pragmatic issues in conducting a study or the convenience of an investigator, in general inclusion criteria are selected precisely because the investigator feels they are clinically relevant to the question being studied, that is that their presence or absence is thought to likely make a real difference in the outcome of the study. Thus, the burden for saying that they do not matter in any given case rests on the person wanting to dismiss them, and some reasonable argument needs to be made as to why a particular criterion can be disregarded.

The EBM argument here states that these differences are likely to be unimportant because they are generally quantitative differences. It is unclear why this would render them unimportant. Quantitative differences of sufficient magnitude can make a major difference. Otherwise, by this reasoning, it would be appropriate to apply results of a trial in geriatric patients to the treatment of two-year-olds. After all, their ages simply represent a quantitative difference. The further argument that qualitative differences in response by different subgroups of patients to a given treatment are quite rare would appear to be clearly untrue, at least in the case of psychiatry. Finally, EBM states that a qualitative difference should only be taken seriously if it makes biological sense, was hypothesized before the trial and is confirmed by a second independent trial. This applies a higher standard of evidence to these qualitative differences in response than are applied to taking the original trial itself seriously. One wonders, given the implausibility of these arguments, if the advocates of EBM feel compelled to reject the view that trials to which our patient would not have qualified may not be applicable to that patient's treatment because to do so would greatly weaken the applicability of EBM to clinical practice. After all a high percentage, probably the majority, of patients that clinicians see in their practices would, for one reason or another, not qualify for most of the clinical trials that have been conducted on their condition.

2.7.3.2 Is the intervention feasible for this patient?

The second question that EBM asks in its effort to address step four is whether the treatment being considered is feasible in the treatment setting in which the clinician and patient operate. This is a straightforward, and noncontroversial, pragmatic point. One cannot offer a treatment to a patient if it is not available or affordable in that setting or cannot be provided competently.

2.7.3.3 Assessing benefits and harms for this patient

The third question used to address step four is to ask about the individual patient's potential benefits and harms from the therapy being considered. This is a reasonable question, but how can one answer it? EBM offers two approaches: a long way that is presented as more rigorous, but more time-consuming and usually not doable in the absence of certain types of study results, and a short way which is presented as the more usual approach in actual practice.[35] We will forgo looking at the details of the long way as EBM readily acknowledges that in most cases these conditions do not apply. Typically, the second way is used to estimate an individual's patient likelihood for benefits or harms from treatment. It consists of making an estimate of our patient's risk of benefit or harm relative to that of the average control patient and expressed as a decimal fraction called f. This is then multiplied by the overall likelihood of the benefit or harm as it occurred in the patients in the study.

> For example, if we think that our patient (if left untreated) has twice the risk of the outcome as control patients in the trial, f=2; alternatively, if we think our patient is at only half their risk, then f=0.5. We can use past clinical experience and expertise in coming up with a value for f or we can use any of the information sources described [for the long method].[36]

Because the authors have already acknowledged that the necessary information for the long calculation is usually not available, all that is being said is the clinician should use their clinical experience and expertise to come up with the value for variable f. However, this of course provides no guidance whatsoever to the clinician, whose judgement has already been placed at the bottom of the evidentiary hierarchy. There is in fact no methodology being offered for the clinician to work back from the clinical trial to a treatment decision for Ms Jones. The vacuous nature of this position is thinly disguised behind a particularly obvious example of pseudo-quantification. The pseudo-quantification is the assignment of a number such as twice as likely, three times as likely, half as likely, to present a veneer of some real precision or methodological rigour to this process. Clearly it does not. EBM begins by transforming the clinician's specific question about treating Ms Jones to a general question about relative response rates in groups of patients resembling Ms Jones. In its effort to be a comprehensive clinical methodology it realizes the need to bring this back to the treatment of Ms Jones. The individualized clinical application is the goal of step four after all. Unfortunately, step four ultimately accomplishes nothing other than admonishing the clinician that they need to somehow, by some unspecified method that nonetheless yields a specific numerical value, circle back to the individualized treatment of Ms Jones!

2.7.3.4 Integrating the patient's values and expectations

The fourth and final question EBM uses in applying the results of clinical trials to the individual patient is to inquire about our patient's values and expectations for both the outcome we are trying to prevent and the treatment we are offering. This question aims to consider the patient's values and preferences and to incorporate those into a treatment recommendation. This goal is laudable, but again the method proposed is a curious one characterized by pseudo-quantification. It is noteworthy that by the patient's values and preferences the discussion is limited to how much the patient would dread potential adverse outcomes and how highly they value hoped-for positive outcomes. This certainly fails to consider a vast array of other value-laden issues important to patients as they consider treatment options. For instance, psychiatric patients often have strong preferences between psychotherapeutic and pharmacologic treatments. Other kinds of complex patient values relating to self-esteem, dignity, stigma, positive and negative prior treatment experiences, and one's theorizing about the nature of one's own condition are not really considered. All of these are often very important to patients,[37] and that is especially true in the treatment of psychiatric conditions.[38]

Yet the proposed method consists of simply asking the patient to quantify as some number, to two decimal points in fact, between zero and one how badly they view the possible adverse effects of the proposed treatment as well as the ill effects of not treating the disorder. These 'severity factors' (s) are then multiplied by the probabilities of the adverse events or the consequences of not employing the treatment in question for this patient, usually based upon the occurrence rate in the clinical trials—the number needed to harm (NNH) or the number needed to treat (NNT), respectively[39]—and the f factor for this patient. This generates a 'likelihood of being helped versus harmed' or LHH. This number is then presented to the patient in the belief that it will clarify for the patient whether they want to accept or reject treatment. As an example of a patient with multiple sclerosis deciding whether to start interferon treatments, 'This LHH can then be presented to the patient, and he can decide whether it is favorable enough to offset the pain and hassle of having to inject himself with interferon every other day'.[40]

I find it unimaginable that many patients facing a complex and difficult decision about their care will find this calculation and the resulting numerical value of much assistance, let alone anyone with a different cultural background, limited education, intellectual disabilities, depression, anxiety, or psychosis. Furthermore, the emphasis is on generating a numerical value that has no precise quantitative meaning. Instead, concrete guidance for the clinician is needed to aid in the difficult challenge of helping the patient sort through their values and preferences when facing serious illness. EBM acknowledges that there may be important issues here, but has little to offer.

> Often the questions that we and our patients ask relate to their experience of disease, diagnostic tests and treatment rather than their test results or health outcomes. Research that explores these issues by asking patients about, or by observing their experiences of diseases, tests and treatments is called 'qualitative research'. Whilst we regard the integration of qualitative research to be one of the major current challenges to EBM, we readily admit that we are not expert in this field and defer to others.[41]

2.8 If EBM is inadequate as a method for clinical reasoning, what is the alternative?

Given this consideration of EBM, what is a reasonable verdict on its adequacy as a methodology for clinical reasoning and decision-making in medicine in general and psychiatry in particular? I believe the following conclusions are justified:.

1. The original mission of EBM has real value, namely to help busy clinicians access relevant clinical studies and to assist them in critical appraisals of those studies—their strengths and weaknesses.
2. By emphasizing the value of RCTs and data derived from studying groups of patients at the expense of valuing the clinicians understanding of an individual patient, EBM carries the risk of denigrating important aspects of the clinician's task.
3. While EBM claims to offer a methodology to bring evidence from clinical trials back to a meaningful application to an individual patient, this method proves to be quite vacuous and disguised by pseudo-quantification.[42]

In short, EBM does not provide an adequate methodology to guide the psychiatric clinician in making decisions in treating the individual patient. Something else is needed.

To achieve this aim, we may not need to generate an entirely novel methodology. After all, we psychiatrists have been evaluating and treating individual patients for many years. What we do with patients is often reasoned and rational, but the process is generally poorly articulated and taken for granted. The claim I will develop in the following chapters is that the foundation underlying good psychiatric practice turns out to be a definable, rational methodology for treating individual patients. Based on expertise and evidence, it is fully scientific in accordance with our current best understanding of what constitutes scientific method. By articulating this process, we can be reassured of the legitimacy of our clinical work, allowing more critical assessments of our methods, encouraging refinements and improvements, assisting greatly in the teaching of this method to new practitioners entering the field, and providing credibility for the knowledge about our patients that we generate in this process.

I think the best way to develop an understanding and appreciation of this methodology is to first demonstrate it at work with several individual patients. Phenomenological explorations of four cases will be presented in some detail in Chapters 3 through 6.[43] Each case will be discussed first in simple descriptive terms characterizing the issue for which the patient seeks help, the treatment decisions by the psychiatrist and the basic outcome of the case. We will consider to what extent EBM principles were helpful to the clinician in that case. I will then make explicit the clinical reasoning used in each case to illustrate a basic methodology for clinical reasoning in psychiatry. This will allow us, in Chapter 7, to provide a detailed general characterization of a method of clinical reasoning and decision-making in psychiatry that captures what experienced psychiatrists in fact do in working with their patients. Each of these four cases were treated by me with enough details altered to hide the identity of the individuals, and except where noted I am the psychiatrist referred to in each case. The cases are varying in complexity but they are typical of the sorts of patients a psychiatrist in private practice might see. I do not believe my reasoning about cases is particularly unique or idiosyncratic, as I believe that the reasoning is characteristic of what most psychiatrists would employ in treating patients. I am thus spelling out my thinking as representative of typical clinical thinking in psychiatry. As explained more fully later, this does not mean that every psychiatrist would come to the same conclusions about a given patient. Nor am I claiming that at the time of treatment all the clinical reasoning was as consciously articulated in my mind as the discussion after the fact might suggest. Indeed, one of the goals of developing this kind of methodology is to encourage the psychiatrist to more fully articulate one's reasoning at the time of treatment to refine and improve treatment decisions. The strategy of retrospective conscious reconstruction of one's thinking processes is of course not infallible, but it has been shown to yield valid and useful results in qualitative studies of human decision-making and expertise development in a range of fields.[44] A more detailed consideration of such reconstructions is considered in sections 7.2 and 8.13–16.

Endnotes

1. Sackett *et al.* 2000; Straus *et al.* 2019
2. Gupta 2014
3. Bluhm and Borgerson 2011
4. APA Presidential Task Force on Evidenced-based Practice 2006
5. Bluhm and Borgerson 2011, p. 203
6. Sackett *et al.* 2000
7. Straus *et al.* 2019
8. Sackett *et al.* 2000, p. 1
9. Montori and Guyatt 2008; Guyatt *et al.* 2001, p. 7
10. A double-blind randomized controlled clinical trial has the following elements: 1) One or more treatment groups and usually a control group that receives either a placebo or 'treatment as usual'. The outcome of the treatment group is compared with that of the control group, and/or the outcome of multiple treatment groups are compared with one another. 2) A randomization procedure to allocate patients to a treatment group 'by chance'. 3) Ideally a blinding procedure is used so that the patient, the treating clinician, and the person doing outcome assessments are unaware of to which group any given patient is assigned.
11. A naturalistic or observational study involves systematic collection and comparison of outcome data for treatments that occurs in the usual course of clinical care, in which no effort is made to influence how patients are assigned to treatments or to standardize how the treatments are given. The treatments may have already occurred at the time the study is conducted (retrospective) or collected as the treatment occurs (prospective).

12. Straus *et al.* 2019, p. 77
13. Sackett *et al.* 2000, pp. 6–7
14. Sadler 2013; Johns Hopkins School of Public Health 2015
15. Benazzi 2007
16. Berk and Dodd 2005
17. Phelps *et al.* 2008
18. Weisberg 2014
19. Sadler 2021
20. Weisberg 2014
21. Sackett *et al.* 2000, pp. 13–27
22. Pillinger 2021
23. Sadler 2021, p. 366
24. American Psychiatric Association 2013
25. American Psychiatric Association 1980
26. Sackett *et al.* 2000, pp. 118–119
27. Khan and Brown 2001; Sonawalla and Rosenbaum 2002; Rutherford and Tandler 2014
28. Hamilton 1960
29. Beck *et al.* 1961
30. Healy 1997, pp. 97–100
31. Sackett *et al.* 2000, pp. 3–4
32. Kwag *et al.* 2016
33. Straus *et al.* 2019, p. 36
34. Sackett *et al.* 2000, pp. 118–119
35. Straus *et al.* 2019, pp. 104–107
36. Sackett *et al.* 2000, pp. 121–122
37. Ross *et al.* 1993
38. Bethany *et al.* 2010
39. The NNT is an estimate of how many patients need to be treated until one patient experiences a benefit from the treatment. The bigger the effect size of the treatment, the smaller the NNT. Conversely the NNH is an estimate of how many patients need to be treated until one patient experiences an adverse event.
40. Sackett *et al.* 2000, p. 128
41. Sackett *et al.* 2000, p. 21
42. Recently there has begun to emerge some awareness of the inadequacy of traditional EBM methodology to make treatment decisions about individual patients. Some more nuanced strategies have been suggested (see Goldberg and Stahl, 2024). While many of these are quite compatible with the approach of this book, there remains a need for an explicit, comprehensive, and epistemologically justified methodological framework to help clinicians understand their own reasoning and decision-making processes.
43. Three of these case histories of Ms K (Chapter 3), Mr Z (Chapter 4), and Mr D (Chapter 5) are adapted with permission from Heinrichs, D. 2015. In J. Z. Sadler, C. W. Van Staten, and K. W. M. Fulford (eds). *The Oxford Handbook of Psychiatric Ethics*. Oxford, UK: Oxford University Press. In all cases discussed in this volume identifiable details were altered to prevent recognition of patients, without altering the issues relevant to the clinical decision-making process being illustrated.
44. Larkin *et al.* 1980; Klein 1998, pp. 13–14

Chapter 3
The role of models in clinical reasoning
The case of Ms K

3.1 Ms K's initial presentation

This and the next three chapters will each examine a patient treated by me to show the psychiatrist's deliberative process in seeking an understanding of the patient's unique circumstances. The result is the construction of a model for each patient that suggests possible treatment interventions. The fate of those interventions tests the adequacy of the model. From these examples, common principles will emerge to allow us to construct a general methodology of model construction as the core of clinical reasoning and decision-making in psychiatry.

Ms K sought psychiatric treatment for the first time at age 58. She reported very distressing anxiety and worry with trouble concentrating, loss of appetite, insomnia, nausea, tremor, and catastrophic thinking over the preceding two months. She related this to health problems of both of her parents resulting in the need to move them to an assisted living facility. She repeatedly emphasized, with some self-condemnation, that she did not know how to cope with adversity as well as other people because her life had been so ideal until now. She described warm and loving parents and a childhood characterized by emotional and material security. Her parents had always sought to spare her distress as much as possible, at times to an extent that would clearly qualify as overprotective. She did well socially and academically. She became a high school math teacher and has been happily married since age 26. Her two grown children are doing well, although her son had some therapy and perhaps some medication for anxiety as an adult. When her husband started his own business, she quit teaching in her late forties to be his bookkeeper. She missed some rewards of teaching, but realized that she had always felt considerable stress in that role despite doing well. She reported that her mother had always been highly anxious, as had several other maternal relatives. She clearly felt uncomfortable seeing a psychiatrist but desperately wanted help. After discussion with the psychiatrist, Ms K opted for short-term supportive therapy and a selective serotonin

How Psychiatrists Make Decisions. Douglas W. Heinrichs, Oxford University Press. © Oxford University Press (2025).
DOI: 10.1093/9780198960720.003.0003

reuptake inhibitor antidepressant (SSRI) and has done well. The patient felt better on the SSRI than she ever had, and chose to stay on it after therapy ended.

3.2 Going beyond evidenced-based medicine: Developing a POP model

In many ways Ms K represents a straightforward and uncomplicated case. She presented with a discrete complaint, treatment was initiated, and a good result was obtained. Such a case is a good place to begin in asking how the psychiatrist developed an understanding of the patient, made treatment decisions, established a collaboration with the patient, implemented treatment, and assessed the outcome. We can first ask how helpful the evidenced-based medicine (EBM) approach would be in this instance. From the case of Ms K, how would we generate an answerable question? The question should take the form of how helpful potential treatments, pharmacologic or psychological, are in a group of patients like Ms K. Although Ms K has a history of frequent but manageable anxiety, she has throughout her life functioned well with minimal distress. She only would have met the full criteria for generalized anxiety disorder (GAD) in the past two months, and this is too short to assign the diagnosis which requires at least six months[1] So, clinical trials using pharmacologic interventions and/or psychotherapy with GAD do not strictly speaking apply to Ms K. She is most appropriately diagnosed as suffering from an adjustment disorder with anxiety. Clinical trials with this group of patients are extremely limited and provide little guidance. Yet, I think many psychiatrists treating a patient such as Ms K would consider an SSRI and some psychotherapeutic modality. What would be the rationale for this?

I suggest that a psychiatrist in this situation generates a model of interacting factors specific to Ms K posited to have brought about her current symptoms, and then hypothesizes how certain therapeutic interventions would influence those interacting factors in reducing Ms K's distress. That model may be more or less explicit in the psychiatrist's mind, but without some such model the transition from the patient's presentation to a treatment plan would be obscure.[2] Often the starting point is an implicit model that the patient brings to treatment and communicates directly or indirectly in telling one's story to the psychiatrist. Patients naturally ask themselves how their current distress came about and often have some account about this that they report to the psychiatrist. Ms K's own account can be summarized as follows: 'I cannot cope well because I never really had to learn how since I had an unusually favourable life. My parents and later my husband made sure of that and always took such good care of me. But I am ill-prepared to deal with my current stressors'.

3.3 Representing models: Narratives and diagrams

A model is an abstract mental construct, but one can represent or articulate the same model in a wide variety of ways. It can, for example, be linguistically expressed in narrative form or represented diagrammatically. Each approach has its representational strengths and weaknesses. Thus, a diagram often makes it easier to see the key components of the model and how they relate to one another, even as it loses some of the nuance of a detailed linguistic description. Figure 3.1 is my diagrammatic representation of the implicit model that Ms K provided to explain to me the origins of her own current distress and her history and current circumstances, using a box and arrow diagram. Each box contains a proposed propensity, and the arrows

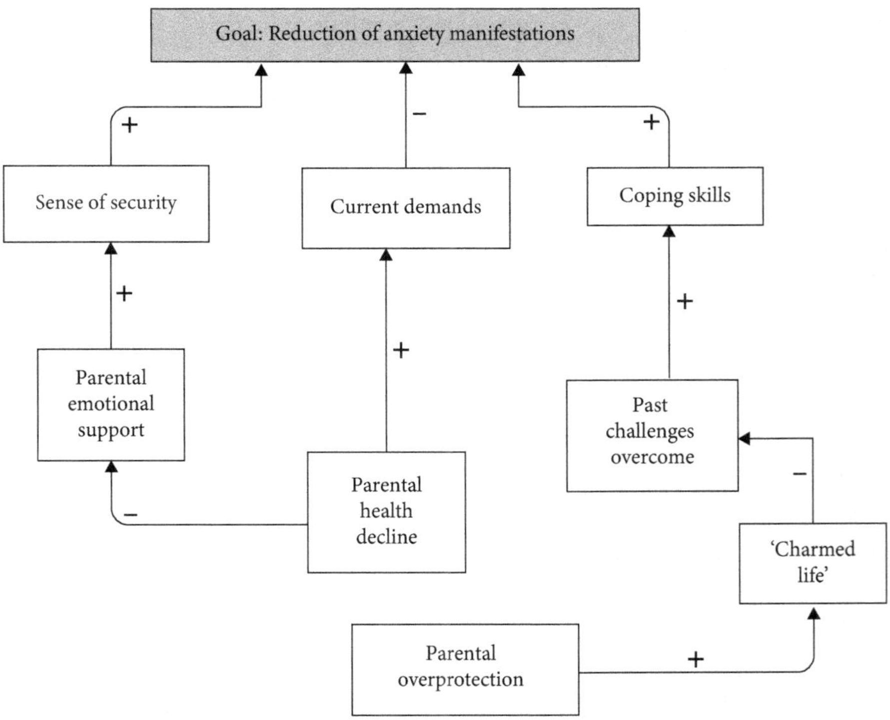

Figure 3.1 Ms K's own model.

Reproduced with permission from Heinrichs, D. 2015. Model-based science and the ethics of ongoing treatment negotiations. In J. Z. Sadler, C. W. Van Staten, and K. W. M. Fulford (eds). *The Oxford Handbook of Psychiatric Ethics*. Oxford, UK: Oxford University Press.

show the interactions between them. Plus and minus signs are used to indicate whether one propensity increases or decreases another propensity, not whether the effect is desirable or undesirable. For instance, the box labelled 'parental health decline, impacts 'current demands'. The plus sign means that parental health decline increases the current demands for the patient, not that it is a good thing. In turn, the arrow from 'current demands' to the box labelled 'Goal: reduction of anxiety' has a minus sign, indicating that these demands work to decrease the reduction of anxiety.

3.4 The clinician's initial model

As I listened to Ms K's story, my own tentative explanatory model began to develop. Again, the modelling process may be explicit and intentional, or implicit. In some cases, the patient's own model will make sense and form the basis of the psychiatrist's own. At other times, a novel, divergent model will emerge that will need to be reconciled with the patient's understanding if a working treatment alliance is to be established. So how did I proceed in the case of Ms K? The thinking process involved draws on and integrates the first two skill sets discussed in Chapter 1; namely knowledge derived from generalizations about groups of patients, in this case from work in psychopathology and therapeutics, as well as an individualized picture of the patient assembled from history taking, mental status examination, and sharing by the patient in the context of a developing therapeutic alliance. The paucity of relevant studies for Ms K's diagnosis makes the EBM strategy of limited value. Yet I considered the more extensive data relating to the treatment of GAD as applicable to Ms K, despite her failure to meet diagnostic criteria by DSM (*Diagnostic and Statistical Manual*) standards. I hypothesized that the set of interacting factors bringing about Ms K's distress resembled some of the factors typically operating in patients with GAD, opening the possibility that treatment trials of GAD may be applicable to her. This strategy is not legitimized in terms of the DSM, which as currently construed is explicitly atheoretical. Hence the diagnostic categories make no claim to any underlying theory or hypothesis to explain them. However, I hypothesized, by way of a model, a pattern of factors influencing the generation of her symptoms. I felt that some intrinsic disposition to high anxiety existed in Ms K based upon her personal and family history, even though that anxiety did not meet a clinical threshold until she experienced the stress of her parents' ageing and health issues. Furthermore, I speculated that the disposition may have some familial basis given the presence of several close relatives who struggled with anxiety. (To what extent this represents genetic

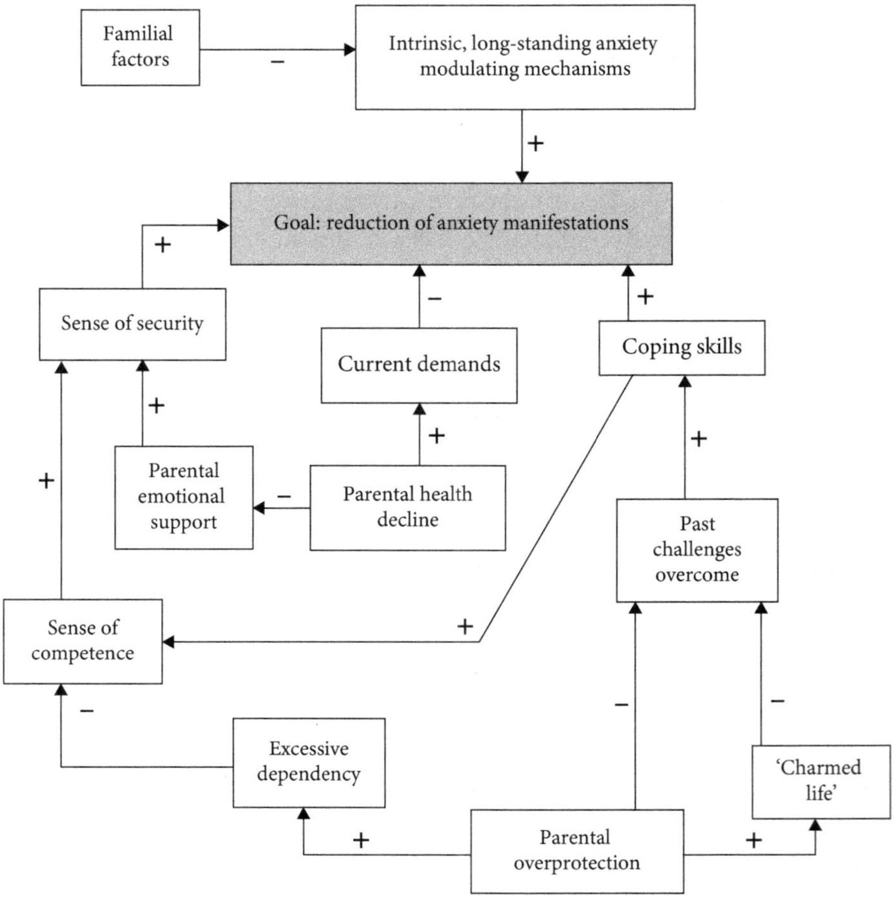

Figure 3.2 Psychiatrist's initial model for Ms K.

Reproduced with permission from Heinrichs, D. 2015. Model-based science and the ethics of ongoing treatment negotiations. In J. Z. Sadler, C. W. Van Staten, and K. W. M. Fulford (eds). *The Oxford Handbook of Psychiatric Ethics*. Oxford, UK: Oxford University Press.

vs. environmental influences was left an open question.) Hence, I developed a provisional model. Note that the model is an imaginative act regarding a specific individual, but one disciplined and informed by the general knowledge base of psychiatry. A diagram that reflects my initial model for Ms K is presented in Figure 3.2.

3.5 Models are provisional

The notion that a model is a provisional representation of an evolving process is central to clinical reasoning. For a model to have credibility beyond that of an educated guess, it must be subjected to ongoing revision in light of new clinical evidence. This evidence can include newly obtained historical information about the patient, subsequent responses to treatment interventions and reactions to new life circumstances. The more a model points to likely sources of future concrete evidence that argues for or against the current model, the stronger and more credible that model is. A corollary of this process is that clinicians could start from a number of diverse initial models of a given patient, but as evidence accumulated over time the revised models would be expected to converge towards a more adequate, accurate, and useful model. However, this process is never complete.[3]

This process is never complete in part because there is always more information about the patient that can come to our awareness. Beyond that, however, for any given body of empirical observations, it is always possible to construct more than one explanatory theory that fits that evidence. In the philosophy of science, this is often termed 'the under-determination of the theory by the evidence'.[4] Hence for any given patient, considering our current body of information, the psychiatrist could construct more than one plausible model. As such, while a good model needs to be compatible with the currently available information about the patient and our general knowledge base, the actual model developed by the clinician is a creative act that draws upon capacities of the imagination. With growing clinical experience, the clinician gains an enriched sense of the imaginative possibilities of what can happen with a patient in response to changing circumstances, including therapeutic interventions. Hence, all models must be viewed as provisional, and the clinician needs to avoid becoming so wedded to any model of a patient that the need to revise or replace it in the light of new evidence is resisted. For a good psychiatrist, imaginative capacity and humility are complementary clinical virtues.

3.6 Some constraints on models

Talk of model construction as a creative, imaginative act does not render it whimsical or even unscientific. There is extensive use of the imagination in the generation of any scientific hypothesis. But there must be considerable constraints on the process. Clinicians are not artists facing a blank canvas who can generate any notion that strikes their fancy. Two important categories of constraints for the clinician are epistemological and ethical.

In positing elements and interactions for a model, the clinician should draw from those that are probable considering our best current knowledge of the factors at work in conditions resembling those of the patient. Thus, for Ms K, my selection of components for my model drew on my understanding of factors often involved in GAD. Early in psychiatrists' careers this sort of knowledge is primarily drawn from their reading of the literature and what they have been taught by supervisors. Over time, the growth of their own clinical experience with the varied patterns of many patients substantially broadens the imaginative possibilities for model construction, thereby facilitating the construction of more individualized models unique to specific patients. One of the problems with a dogmatic reliance on the methods of EBM is that it tends to lock clinicians from very early in their careers into the base of information available to them, and discourage them from exploring the enhanced imaginative possibilities for understanding the patient that comes with their growing experience.

I started with Ms. K in part because she is an example of a straightforward clinical situation that a beginning psychiatrist or resident can understand with the available knowledge base. Some of the subsequent cases we will consider draw more extensively on possibilities for understanding that depended more upon my own subsequent clinical experience. Yet even for a case like Ms. K, elements of the model, and especially potential treatment options, will vary based upon where and how one trains. Having come from a strongly psychodynamic training programme, the most salient treatment options for me will be different from a psychiatrist with a strong cognitive-behavioural background. This is not in and of itself a problem. There is no reason to assume that there is only one correct model. It is likely that there are several models that could bring about the desired benefit for the patient.

When we construct a model for a patient that will guide our treatment decisions, beyond such epistemological constraints, there are ethical constraints as well. A model is stronger to the extent that it generates predictions of what will happen with possible interventions. But our patients are not experiments in a test tube. It is incumbent upon us to consider the potential consequences if the model is wrong. Are we proposing an intervention that might considerably harm our patient? Thus, we must imaginatively project the ways in which our model can be in error as well as avoiding interventions that carry an unacceptable level of risk. As a result, we must prune away from the range of potential interventions suggested by the model any that pose unacceptable risks. We must temper our enthusiasm for our ideas with prudence and humility. The modern history of psychiatry is unfortunately replete with enthusiastic innovators who failed to do so with sometimes disastrous consequences, from random removal of organs thought to be infected to prefrontal lobotomies.[5]

3.7 Models as patterns of propensities (POP)

Regarding Ms K, included among the elements in my initial model is an intrinsic and longstanding propensity to anxiety interacting with recent life events. I further understood that interaction in terms of strong dependency needs, probably resulting both as a reaction to her intrinsic anxiety and the particularly overprotective caretaking style of her parents. The model consisted of several elements connected by paths of influence between them. One might call this a proposed mechanism for the generation of the patient's distress. For reasons that will be only fully appreciated in the later discussion in Chapter 10 of the nature of scientific method, 'mechanism' can be a problematic term. Suffice it to say here that it is often used to describe a system of causal, deterministically interacting elements in a highly circumscribed entity. Given the controversies about the nature of causality and the degree of strict determinism in psychological processes, as well as the limited ability to make precise predictions in complex systems and especially those involving human behaviour, I think that different terminology is preferable.

Given that human beings are not closed systems, the influence between elements can never be predicted with certainty due to the potential impact of many other extraneous conditions. One can only say that certain factors have a tendency or disposition to influence other factors, or the capacity to bring about certain consequences, all things being equal. To keep that uncertainty in focus, I think the elements involved can best be called **propensities**. The stronger case for the appropriateness of that term will be made in sections 7.7, 7.8, and 10.2.3.8. The term propensity, then, is used to denote an element or state of affairs that, when present in a system of interacting elements, has a disposition to bring about certain results. Whether those results in fact occur, and to what degree, is dependent upon the context of the other elements present. For instance, the concept of gravity defines the propensity of two bodies with mass to move towards one another. In any real system of two such bodies, whether they in fact move towards one another and at what velocity, depend upon the array of other elements and forces in the field. The standard force of gravity ($F=G\ Mm/r^2$) is a theoretical construct of how two bodies would behave in a situation where no other objects or forces are present. Strictly speaking this never occurs in any real-life situation. Hence that formula defines the strength of the propensity of gravity that will always be altered to some degree by the context of other forces and elements. The context will be vastly more complicated, with many more components and forces at work, in a model intended to describe the feelings and behaviour of a human being. Thus, my model of Ms K hypothesizes some familial propensity to anxiety. Whether and to what extent it manifests itself in any given circumstance will depend on the context.

In constructing a model, then, I hypothesized a group of interacting propensities that I thought were relevant to generating the patient's distress. I further speculated on the pathways of interaction between these propensities to develop the model as to how her distress came about. So, instead of referring to this sort of model as a mechanism, I call it a **pattern of propensities (POP).**

3.8 Models are organized around goals

Several additional features of the model become clear from examining the diagrams. The POP is not simply some unspecified model of Ms K. Rather it is organized around a particular goal, in this case the reduction of her anxiety and accompanying distress. A model only has meaning in relation to a purpose the model is intended to serve—the problem it is intended to solve. As will be discussed at length in section 10.2.3.3, this is true of all modelling in any science. There are many things in terms of which one could construct a model about Ms K. I was specifically concerned with the distress that brought her to treatment and how to alleviate her distress. There is no such thing as a disinterested model. It always exists to serve some purpose and its adequacy can only be judged by its success in achieving that purpose.

A useful way to think about how models work is using an analogy of a map.[6] Maps are in fact models. A map is designed to represent a particular piece of the world. No map can capture every aspect of that piece of world. Many maps can be created for the same piece of the world, mirroring various specific aspects of that piece with varying degrees of accuracy. What makes a particular map a good map cannot be defined without knowing the purpose of the user. A roadmap must reflect the relationship between locations and the roads that connect them. It often is drawn to scale, but need not be in all cases to work adequately. It typically does not reflect topography as this is not usually important if one is driving a car. But if the map is being used by a hiker, topography might be quite important.

Many features of the map are highly inaccurate as representations of that piece of world. In a roadmap all cities may be represented by simple dots. Obviously, there is no physical resemblance between the city and that dot, but for the purposes of the map user that is not important. Furthermore, the features that the map does choose to represent will never be perfectly accurate.

How accurate they should be to be a good map depends completely on the purpose of the map. If a friend is drawing a quick map to help me get to a particular location, I do not expect precision in scale or precise orientation. Usually all I need is the relative relationship between where I start and where I want to end up, and directions at relevant decision points along the way. If the map gets me there it is a good map. It serves its purpose. Many of these issues relating to maps apply directly to models in general and POPs in particular. What elements need to be included and how precisely they need to be measured and characterized depends totally on the goal we wish to achieve using them.

3.9 Initial uses of the model

There are initially two functions of our POP for Ms K. The first is to employ it to establish a shared goal and a collaborative understanding that makes sense to the patient as to what factors may have brought about her distress. In this case my initial discussion with the patient, highly condensed, was essentially as follows:

> It sounds like you have indeed had a fortunate life. But there is a lot of research suggesting that the propensity to anxiety also has strong familial contributions. Your son has been treated for it and your mother and several of her relatives have struggled with it chronically. I suspect that you have virtually always had a significant propensity to anxiety yourself: consider for instance the stress you felt while teaching even though you functioned very well. That also shows that you have some capacity to cope with stress. My understanding of what is happening is that the current demands of dealing with your parents' declining health, combined with the loss of the security they have always provided, has led to a buildup of stress that has exposed your underlying propensity to anxiety. A supportive and nurturing environment, both as a child and adult, protected you from clinically significant anxiety before now, but also may make you feel especially reliant on key persons in your life rather than on yourself as sources of strength.

This presentation of the model made sense to Ms K. The next use of the model is to generate proposed treatment options. To do this I used the model to look for points at which various treatment options could influence either the propensities themselves or the strength of the connections between. In other words, the model itself is expanded to include new propensities that are introduced by treatment. Figure 3.3 represents this expanded model. For

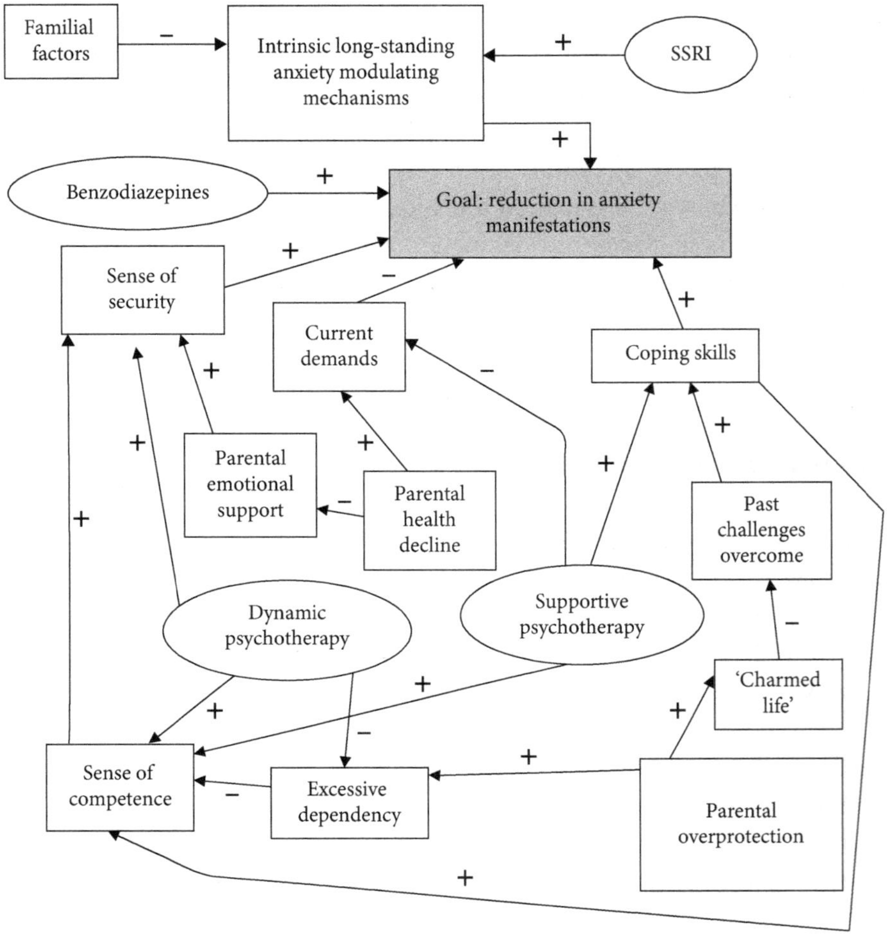

Figure 3.3 Psychiatrist's expanded model for Ms K.

Reproduced with permission from Heinrichs, D. 2015. Model-based science and the ethics of ongoing treatment negotiations. In J. Z. Sadler, C. W. Van Staten, and K. W. M. Fulford (eds). *The Oxford Handbook of Psychiatric Ethics*. Oxford, UK: Oxford University Press.

the sake of clarity, the treatment options are presented in circles rather than squares. Three treatment options are considered: SSRIs, benzodiazepines, and psychotherapy.

The selection of possible treatment options reveals an important aspect of model use by clinicians. The hypothesized model helps to suggest potential points of intervention in the POP. Again, remember that the plus or minus signs accompanying the arrow leading from a potential intervention to a propensity in the model is meant to indicate only whether the intervention increases or decreases the strength of that propensity, not whether that is a desirable or undesirable effect in and of itself. We can only determine whether it is desirable by tracking the path from the proposed intervention through to a goal. If it strengthens the goal it is hypothesized to be desirable, and if it weakens the goal it is undesirable. But there is a wide range of interventions that could be considered. Which of them should be selected? Throughout the discussion of these cases the role of patients' preferences and values will be frequently considered. Here I want to emphasize the importance of the clinicians' unique backgrounds, competencies, and how they are acculturated into their professional roles. Different clinicians may well propose different treatment interventions for the same patient. This need not be a problem so long as there is a reasonable basis for the plausibility of the interventions, a careful consideration of potential adverse effects, ongoing assessment of the effectiveness of the interventions employed, and a willingness to change interventions in the light of treatment failure. There is every reason to assume that more than one mix of interventions could prove helpful and effective for a given patient. A POP model makes no claim to result in some absolute optimal treatment. Rather it is designed to lead to an adequate treatment that sufficiently resolves the patient's difficulties. We never really know whether some even better treatment exists for a given patient.

To avoid getting distracted with a potentially endless series of discussions as to the relative merit of alternative interventions to those I proposed to the patients in these case presentations, I will limit myself to presenting a basic rationale for my proposed interventions sufficient to illustrate the use of models in my clinical thinking. In fairness I will state my own background here. My training emphasized both psychopharmacology and psychotherapy and encouraged their combined use in most cases. Having been trained in the 1970s, my psychotherapy competence is primarily in psychodynamic psychotherapy, although I at times incorporate elements of other approaches such as cognitive-behavioural therapy, dialectical behavioural therapy, and interpersonal therapy. The influence of this background will be apparent as I lay out my proposed interventions with patients and in how I tend to conceptualize issues with patients in general.

3.10 Some treatment options for Ms K: Presentation and negotiation

My further presentation of the model including treatment options can be summarized narratively as follows:

> As to treatment I would first recommend psychotherapy that would look at your lifelong tendency to depend on others over relying on your own efforts in demanding situations in order to help you develop confidence in your own capacities to cope as an adult, as well as addressing the immediate stressors involved with current situation. Dealing with this issue would prepare you to cope with future challenges and losses that are bound to occur. I think there is also a role for medication. Diagnostically, you have an adjustment disorder with anxiety. Some psychiatrists would recommend little medication except something short-term if needed, such as a benzodiazepine, for when the anxiety is overwhelming. However, I suspect it would be best to use a selective serotonin reuptake inhibitor, because I suspect you have the longstanding tendency to anxiety typical of a GAD, even though, due to an especially favourable environment, anxiety was not particularly problematic for most of your life. SSRIs are usually favoured for long-term use over benzodiazepines because the latter are more likely to generate long-term side effects, particularly as we get older. Once you are over this crisis, we can discuss whether the SSRI makes life sufficiently more comfortable than your usual baseline before the crisis to merit continuing with it.

Ms K felt that the notion of a lifetime propensity to anxiety made sense to her and she agreed to start an SSRI. As to the psychotherapy option she was ambivalent. She was eager to get some help in dealing with her current problems but did not feel that she had the emotional strength to look at the past. In addition, she felt that the notion of exploring her dependency issues with her parents in some way represented criticism of them, and hence was uncomfortable with any exploratory psychotherapy. The final treatment decision then was to use an SSRI and some brief supportive psychotherapy focusing on helping her manage her current stressors.

This point in the process is a good illustration of the role of the clinician's particular theoretical and training background. A psychiatrist with a different expertise might well have suggested a course of cognitive-behavioural therapy (CBT). While I incorporated some elements of CBT in my supportive therapy with Ms K, it is an open question how much the process and outcome would have been different with a course of formal CBT and whether Ms K would have found the rationale for CBT more acceptable to her than that of psychodynamic psychotherapy. Nonetheless, the negotiation of treatment with Ms. K in essence represents a revision of the model into something that was acceptable to the patient and that we both could embrace. This new shared model can be seen diagrammatically in Figure 3.4. The use of SSRIs remains in the model while the option for benzodiazepines has been dropped, reflecting the chronic nature of her anxiety. More significantly, the

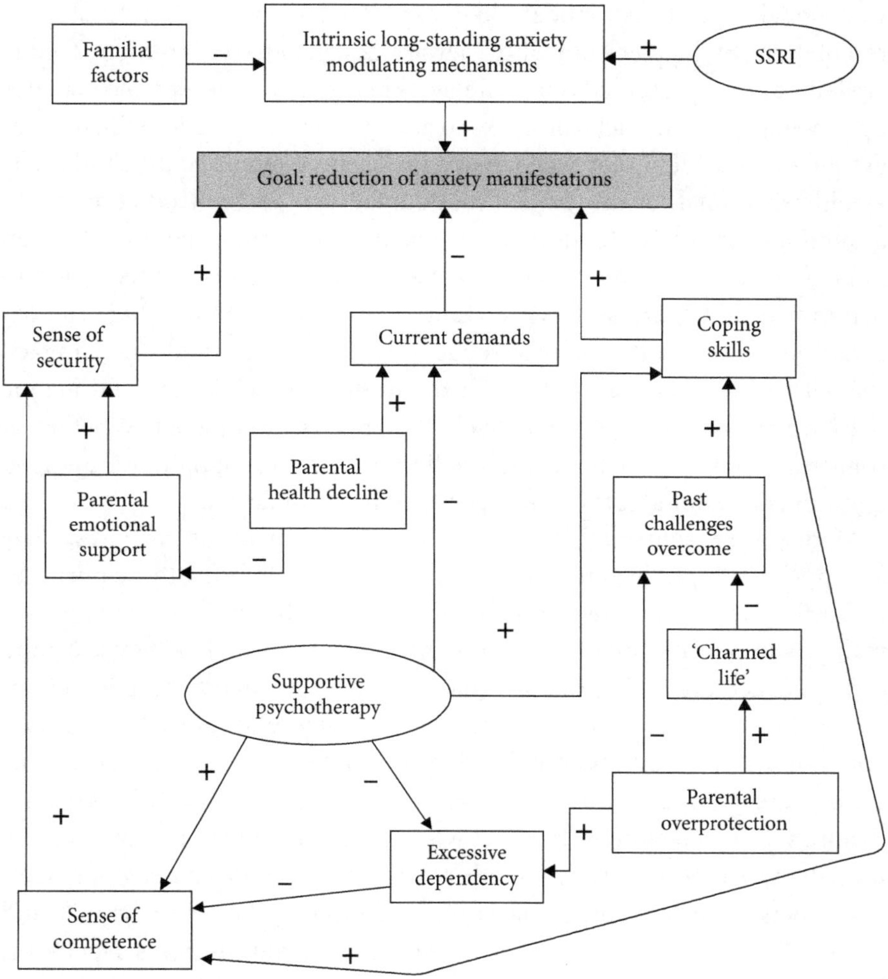

Figure 3.4 Shared model for Ms K.

Reproduced with permission from Heinrichs, D. 2015. Model-based science and the ethics of ongoing treatment negotiations. In J. Z. Sadler, C. W. Van Staten, and K. W. M. Fulford (eds). *The Oxford Handbook of Psychiatric Ethics*. Oxford, UK: Oxford University Press.

role of psychotherapy is changed and has been truncated to include only supportive attention to current issues and coping strategies.

3.11 Models are judged by their utility

This revised model, while acceptable to the patient, does not represent my initial formulation of optimal treatment. Is this model then in some sense false? That way of phrasing the question reflects a misunderstanding of the way models operate. All models, like maps, are less than fully accurate representations of the piece of reality they are designed to reflect. Many factors are left out or represented only crudely. Whether that is acceptable depends upon whether the model still works to achieve the function for which it was designed. So, while I may well believe that a more intensive psychotherapy would be optimal for this patient, undoubtedly in part reflecting my background and training, the question is whether this truncated model, when implemented, will bring about the patient's goal which is the reduction of her anxiety and distress. This is an empirical question. In the short run Ms K responded well with a marked reduction in her anxiety and stress. In fact, she felt less anxious than she ever had and therefore chose to continue the SSRI treatment after the brief psychotherapy was completed. Whether at some time in the future her anxiety will return in a troubling way is an open question, and should it the model can then be revised again.

Models are pragmatic things designed to serve particular purposes. They are appropriately judged in terms of whether they help bring those purposes to fruition. How accurately and comprehensively they reflect some piece of reality is important only from the point of view of whether they are good enough to achieve their goals. This has many parallels in other aspects of scientific endeavour, as will be discussed more fully in sections 10.2.3.3–7. If one is modelling movement of the planets about the sun, for instance, many features are ignored such as the shape, colour, and composition of the planets themselves as well as other forces at work beyond gravity that may influence the pattern of motion, however modestly. The classic model based upon treating planets as single points of mass only influenced by the force of gravity and acceleration turns out to be quite good enough for many purposes. Hence this is a good model for those purposes despite the many omissions of features of the concrete reality of the solar system.

By focusing on the interaction of propensities, the POP model approach attempts to acknowledge the real uncertainty present in understanding the complexity of an individual person. It attempts to capture in an explicit way the actual clinical reasoning of a competent psychiatrist. The POP approach provides a framework for understanding that can be presented to patients with commonsense appeal. A formulation which is transparent to the patient is critical for developing a shared collaboration for the goals and processes of treatment. The POP formulation eschews any pseudo-quantification as discussed in section 2.6. It certainly posits a direction of change, namely that a particular propensity is likely to increase or decrease some other propensity or an observable outcome, but leaves the exact magnitude of that effect unspecified. This accurately reflects the real limits of our knowledge and our limits in making predictions of response. As part of constructing and using a POP model, the clinician undoubtedly has opinions as to which propensities are likely to be relatively more important in each situation, which will be additive with others, and which will serve to nullify the effects of others. This allows the clinician to help hypothesize that certain interventions will bring about the desired changes in the patient's condition and that others will not, or perhaps even make matters worse. This is as good as we can hope for. The case of Mr Z in Chapter 4 involves an example of an especially challenging set of such considerations.

3.12 The importance of pruning models

In the case of Ms K, the negotiation process led to what constitutes a pruning of my original model. Pruning is an important concept in the use of POP models, and the basis for such pruning can be grouped into several heuristically useful categories that need not be mutually exclusive. A negotiated pruning is one that results from a dialogue with patients in which their preferences and level of comfort with various possible interventions lead to the removal of certain potential interventions from consideration for them. We have seen that with Ms K, and it will play an important role in other cases we consider such as Mr Z in Chapter 4, and especially prominently with Mr D in Chapter 5. There are also cases of pragmatic pruning, where otherwise potentially useful interventions are taken off the table because of practical obstacles to their implementation. For instance, a specialized psychotherapeutic approach that might be useful might be impractical because the psychiatrist lacks the skill set to provide it and no other competent practitioners of that technique are available in the area. Or a particular medication that might be deemed of potential value may not be on the patient's drug formulary and be otherwise unaffordable.

An especially important form of pruning is ethical pruning. This occurs where a reflection on a potential intervention results in a judgement that implement of that intervention would be ethically problematic. A common form of this sort of pruning occurs when the potential intervention carries a serious risk of doing harm to the patient. This harm can take two general forms. In the first, even if the model proves to be essentially correct, the intervention carries an unacceptable risk. It is incumbent upon the psychiatrist to consider every intervention that flows from the model as to the expectable risks and to discuss these fully with the patient. The simplest cases are potential medical or behavioural side effects from pharmacologic or other biologic interventions. But it is important to remember that there can be significant adverse effects from some psychotherapeutic interventions as well. Indeed, the negotiated pruning in the case of Ms K can also be viewed as an instance of ethical pruning. Ms K's concern with exploratory psychotherapy included the belief that she was not psychologically strong enough to deal with the emotional turmoil that might result. Whether she was right in this regard or not, it is important to consider this possibility of harm. And such considerations of potential harm need to be assessed by the psychiatrist before implementing the model whether the issue has been raised by the patient or not. After all it is our mandate to first do no harm.

Another pathway to potential harm that the psychiatrist has an ethical obligation to consider is the consequences if the model of the patient is in some way wrong. If a model of a specific patient indicates that an antidepressant is indicated, what is the likelihood that an underlying and unrecognized bipolar process could lead to a worsening of the patient's condition in response to an antidepressant? Or if the model of a patient suggests a role for extensive exploratory psychotherapy, how likely is it that an underlying and unappreciated fragility in the patient could lead to a psychotic decompensation in response to the psychotherapy process? It is ethically incumbent on the psychiatrist to consider these kinds of possibilities before moving ahead to implement interventions based on a POP model.

There are, of course, other possible reasons for ethical pruning of a model apart from concerns about clinical harm. The psychiatrist must be sure that proposed interventions do not violate patients' autonomy, right to informed consent, confidentiality, and dignity. All forms of pruning require a process in which the psychiatrist considers a wide range of 'what if' scenarios and counterfactuals. As such it represents another instance where the imaginative capacities of psychiatrists, honed by their training and accumulated clinical experience, must be brought to bear.

3.13 Illusions of certainty and the virtue of intellectual humility

As conscientious psychiatrists attempting to practice with rigour and intellectual honesty, how satisfied should we be with a POP model of the sort presented here for Ms. K? We would all like more certainty and precision in our knowledge of patients and in our prediction of treatment response. However, we should not pretend that we have more precise knowledge than we really do. Part of the appeal of much of contemporary clinical research may be that it gives the illusion of more precision and certainty than actually exists. On the one hand the precisely characterized methodology may seem to imply that if randomized controlled trial (RCT) methodology is applied, true and accurate results will follow. The clinician need only determine if the study met rigorous methodological standards and, if so, accept the results. This sense of certainty is augmented by the aura of statistical rigour, often poorly understood by clinicians. As a result, clinicians tend to accept statistical analyses on faith as if the application of statistical methods guarantees the quality of the study. Often forgotten are the many assumptions made with statistical methods and that many of these assumptions are only approximately met, if at all, in actual clinical trials or the real world of clinical practice.[7]

Furthermore, there is almost a magical quality to the notion of statistical significance. A p value of $< .05$ is often interpreted to mean that the results can be taken to the bank. But is this so? Such a p value is often taken to mean that there is less than a one in 20 chance that the hypothesized treatment is not effective. What it actually means is that there is less than a one in 20 chance that the actual results obtained in the trial would occur if the null hypothesis is true, usually that the treatment has no effect. This is quite a different matter. Furthermore, when one considers the vast multitude of clinical trials conducted in psychiatry the notion that as many as one in 20 will show statistical significance by mere chance is sobering.

Beyond this, the methodologies of RCTs are meant to eliminate a few types of potential bias, specifically biases in patient assignment to treatments within the study and, if double blinded, to perceptual biases about expected outcome by the treater, the patient, and/or the researcher. We should remember that many other forms of bias are not automatically controlled for by RCT methodologies[8]: what questions are selected for study, what sampling techniques are used to draw the study sample out of a larger population, what differences may exist from one study centre to another, publication biases that favour the publication of positive over negative results, biases in the interests of the investigator either relating to monetary benefit or career advancement, and then the whole host of issues related to problems of external validity as discussed in section 2.6. I am not arguing that clinical trials are of no value, only to illustrate that uncertainty exists in every aspect of the field. Uncertainty is intrinsic to the complexity of the problems we attempt to address. So, the presence of uncertainty in POP model construction with individual patients does not imply that it lacks rigour.

3.14 The power of prediction and pattern recognition

A further safeguard in using POPs follows from their purpose in generating predictions of treatment response. Hence their utility is testable in a specific application in a unique patient. Granted that if the patient responds well, it could have little or nothing to do with the treatments employed or the model used in generating those treatment options. In such circumstances the worst-case scenario is that patients come for help with their distress, they leave us relieved of their distress, and we incorrectly credit our treatment interventions with fixing the problem. While not ideal, this outcome is not the worst one. However, there are factors likely to add some clarity to this matter. If a medication regimen is recommended in an ongoing way to maintain the patient's improvement and the patient strongly wishes to find out whether the treatment in fact brought about the improvement, the treatment can be discontinued and the patient's outcome assessed. This can be repeated multiple times to give an increasing sense of certainty to the result. Not that every patient would choose to run this experiment but it remains an option. This discontinuation-testing may be important in cases where the ongoing treatment carries some potential for adverse effects over time, requiring an assessment of the risk/benefit equation. An extensive and rigorous methodology exists as to how to conduct and evaluate such N-of-1 research.[9]

Key elements in generating a POP model for a patient are (1) a search for patterns of recurrent subjective states and external behaviour, along with (2) identifying the sorts of situations that generate them. For many of our patients the problems that they bring to us will not occur just once in their lifetime. In many cases they recur with some regularity. Each occurrence and each attempted treatment are separate trials. Each positive response adds some cumulative weight to the hypothesis that the treatment is effective. One of the advantages of working with a patient over an extended period is the opportunity to accumulate such instances. Even when the current treating psychiatrist has not worked with a given patient for a long period of time, often prior experiences with their condition and response to treatment can be obtained from careful history taking and record review.

Furthermore, in the alternative scenario where the patient does not improve, we can say with greater certainty that some aspect of our POP is incorrect or incomplete. Often, feeling certain is easier with a negative outcome—the POP is wrong in some way—than feeling certain as to its rightness in the face of a positive outcome. This mirrors a well-known theory in the philosophy of science popularized by the philosopher Karl Popper that what distinguishes scientific thinking is that the hypotheses generated are falsifiable.[10] Popper would maintain that we can never be sure any scientific theory is correct regardless of how much supporting evidence accumulates. He is often taken to further claim that a single concrete occurrence that contradicts the theory proves it false. This view is generally seen as excessively draconian in that a failed prediction could result from inaccuracies in the many auxiliary assumptions made in any observational situation or from errors in measurement or instrumentation. A fairer reading of Popper is that a failed prediction means that something in our understanding or assessment of the situation needs modification.[11]

3.15 Pattern recognition in the domain of the individual

This issue of the recurrence of certain types of clinical problems merits some further discussion, as it is not only relevant to how much confidence we come to have in the model, but also plays an important role in the methodology and rationale behind the construction of the model itself. When we say that our model is an attempt to define a pattern of propensities relevant to the production of the particular occurrence of distress for the patient, how do we begin to hone in upon which elements are likely to be relevant? In selecting from the patient's narrative those components that should be incorporated into our model, one of the things we do is to identify and focus on recurrent patterns relating to similar types of instances in the patient's life. In a very high percentage of our patients the specific distress that currently brings them to treatment is mirrored in a pattern of occurrences and responses across their life.

In the case of Ms K, responses of heightened anxiety were characteristic of her over a wide range of situations and aspects of her life—in her dealings with family, friends, school, and career. As she told her story, this pattern was revealed both by the reporting of numerous examples and through spontaneous generalizations that the patient made about herself. Thus, she described herself as somebody who always tended to get anxious when confronted with anything new or demanding. She described herself as always being security-conscious and intensely aware of the comfort she found from seeing her parents as loving and protective of her. I would argue that in my cumulative experience in evaluating and treating patients, most patients reveal such regularities. I look for them with a new patient, and if they are not spontaneously revealed will ask about them. As clinicians we typically ask about similar episodes of the patient's distress in the past, of both clinical and subclinical levels of severity. We also ask about other factors that we postulate, based on our experience with other patients as well as drawing on the existing literature, are likely to occur in individuals with this sort of pattern of distress. Thus, if a patient presents with depression, we will ask about prior episodes of depression as well as family history suggesting depressive illness or early life factors known to be associated with depression, such as early parental loss.

In fact, we so expect to see repeating patterns that their apparent absence is itself highly significant in directing us to consider certain pattern of propensities. For example, if a 60-year-old individual presents with psychosis for the first time in the absence of any prior clinical or subclinical episodes and in the absence of a family history, the possible role of a definable neurological condition or a systemic medical disease in contributing to the presenting symptoms becomes an important part of our model. Similarly, if a middle-aged person with a lifelong pattern of good psychological resiliency and the absence of significant dysphoria or anxiety becomes depressed in response to a non-catastrophic life event, we are likely to explore the possibility that the current triggering event symbolically resonated with some critical psychological theme in the patient's life, or that an underlying organic explanation exists.

All this is to remind us that when we say we are presenting a model of Ms K, what we are saying is that the piece of the world being modelled is Ms K. She is the domain of our investigation. Within that domain, the model is not an attempt to globally represent Ms K. Rather it models the pattern of propensities relevant to the incident of distress that brings her to treatment. While there is only one Ms K, it is likely that the incident of distress represents one instance of a pattern of occurrences in that piece of world that is Ms K. I would argue that the search for such patterns of recurrences constitutes a key element in the rational methodology of model construction. This distinction between Mrs K as a singular and unique domain and the repeated occurrences of similar types of experience and behaviour over time in that domain, demystifies the apparent paradox of speaking of pattern recognition in an individual, unique patient.

How much confidence should we have in a model constructed in this manner? This issue will be addressed in considerably more detail when we look at the epistemological status of model construction and its relationship to scientific method in Chapter 10. At this point a few observations may be useful.

The staunch supporter of statistical methods might well argue here that even in discovering a pattern in Ms K, we are dealing with a relatively small number of occurrences. Hence predictions based on them would merit relatively low confidence compared to those based on results of large clinical trials. This critique relates to the point made in section 2.4 in discussing Robert Weisberg's elaboration of two meanings of probability.[12] If one is only looking at mathematically generated statistical probability, this critique is undoubtedly true. However, Weisberg identified a historically much older and still valuable process of assessing probability that we use in most of our everyday dealings with the world, and is very explicitly employed in the legal system to determine guilt and culpability. Consider the following hypothetical scenario. You are a juror in a burglary case. A's home was burglarized. B and C were seen near the scene at the time of the burglary. Both B and C have prior histories of two burglaries each. A's home was vandalized with distinctive graffiti painted on the walls. Nearly identical graffiti had been painted on the walls of the two homes that B had previously burglarized. This fact had never been widely publicized at the time of the prior crimes and there is no reason to think that C was aware of this pattern in B's previous behaviour. The prosecutor argues that the presence of the graffiti makes it highly probable that B committed the crime. These facts are not subject to any statistical test of probability or the question of whether a level of statistical significance has been reached. Nonetheless, I suspect most of us would see this pattern as having considerable force as we consider the probability of whether B is guilty. I would argue, and I have every reason to think Robert Weisberg would agree, that this is a totally reasonable and rational way to think.[13] (Whether it reaches a level of beyond a reasonable doubt is a judgement that would also depend upon what other evidence is available.)

Why should this sort of argument, based upon a pattern with multiple but not a vast number of instances, be devalued in clinical reasoning just because it cannot be subjected to standard statistical methods? There is something wrong with a view of scientific method that discourages the use of reasonable and rational ways of thinking that carry significant weight in a myriad of other human decision-making situations.

3.16 Theoretical and training biases in model construction

In sum, within the realistic limits of our knowledge, a POP model articulates a disciplined methodology for clinical reasoning for the practising psychiatrist. In many ways it reflects the kind of deliberative strategies that conscientious psychiatrists have always employed. But by raising this to a conscious and well-articulated method, psychiatrists can be more critical in their self-assessment and can judge more accurately the adequacy of their understanding of the patient, especially when things do not go as predicted. The POP also provides a framework for understanding that can be communicated easily to the patient and form the basis of a collaborative treatment relationship that respects the values and preferences of the patient. Like any other human endeavour, this process is subject to certain biases. Making this process conscious and being aware of possible biases can reduce their likelihood and their impact.

One inevitable sort of systemic bias, not necessarily a problematic one, should be specifically re-emphasized here. The specific propensities and their interactions that a particular psychiatrist assigns to a model of a patient will heavily reflect the training and theoretical frameworks typically employed by that psychiatrist. As I noted in the introduction, my training and background included strong emphasis on psychodynamic conceptualizations as well as developments in psychopharmacology. The reader will, unsurprisingly, note that elements from this background play prominent roles in the specific models I generated in the cases under consideration. Clinicians with different training histories and theoretical orientations are likely to generate models that reflect their specific backgrounds. Should we see that as a problem? Not necessarily. Often differences between formulated explanations from different theoretical positions primarily reflect differences in language over substance. Furthermore, even where there are substantive differences in a model's content, there is no reason to assume that there is only one model that would generate effective treatment for the patient's distress. Remember that a model is a pragmatic thing designed to solve a particular real-life problem, in our case to achieve the patient's goal of alleviating distress. Whatever model the clinician generates should be judged in terms of its ability to meet this goal.

The philosopher of science Nancy Cartwright[14] has pointed out that even in fields like physics there are frequently competing and theoretically incompatible models proposed to address the same problem, such as those using methods from classical versus quantum physics. In some cases, one model is more effective at solving a specific problem than another while at other times multiple models can lead to equally effective solutions. A consequence of this is that the method of POP construction does not imply that all clinicians should generate the same model of given patient. There is no singularly correct answer. What is being proposed is that the use of POP modelling as a general methodology employed by experienced and competent psychiatrists is highly likely to generate useful and pragmatically adequate models for specific patients. As we will see in other examples, it also provides a methodology for modifying and correcting a model that is not up to the task of meeting the patient's goals.

3.17 Applying generalized knowledge in EBM versus POP models

Can the psychiatrist, then, start anywhere in generating a model and simply rely on future modifications based upon successful and unsuccessful predictions to self-correct the model? Such an arbitrary practice would raise serious ethical concerns. After all, treatment interventions are not disinterested experiments. They profoundly affect, for better or worse, the lives and well-being of our patients. Our initial models need to represent conscientious attempts to be helpful and effective, and as much as possible to do no harm. While there are variations in professional background and training, all models need to be rooted in the best available knowledge in the field.

But there is a fundamental difference in how that knowledge is applied in POP modelling in contrast to EBM approaches. The latter argues, for instance, that the result of a set of clinical trials can rather directly tell us what is the optimal treatment for our unique patient. The contact between generalized knowledge in the field and the individual patient is more fine-grained in a model-based approach. Here the role of generalized knowledge is to provide evidentiary support for a particular pathway of influence hypothesized in the model of the patient. Thus, for my model of Ms K, one can ask questions like the following: What evidence is there for familial factors playing a role in anxiety disorders? What reasons do we have for positing that parental overprotection can lead to excessive dependency? On what basis do we assume that SSRIs are likely to reduce anxiety or that psychotherapy can increase coping skills? How confident should we be that increased coping will lead to an enhanced sense of competence and security that will reduce anxiety?

Various kinds of evidence can be brought to bear on each of these questions. Formal results of clinical research certainly count. So do the cumulative understanding from naturalistic observations, both formal and informal, provided both in published literature and communicated to younger psychiatrists by their supervisors and mentors. Insights from various theoretical frameworks and even folk psychology can have a role. Over time the experienced psychiatrist generates generalizations from their own experience base as well. A more detailed consideration of the range of sources of evidence in psychiatry will be discussed in section 9.15. The point I wish to emphasize here is that the initial model formulated by the psychiatrist should not be just pulled from the air or based on some hunch. Rather psychiatrists are ethically obligated, given the potentially serious consequences of interventions, to ask themselves how strong the support for each pathway in the model is. While there are multiple effective models that could be employed, there are many more potentially disastrous ones. This fine-grained application of evidence is also useful in deciding how to modify a model that has failed in a prediction. What are the weakest evidentiary links in the proposed pathways of interaction between propensities in the model that could explain the failure?

Whatever its limitations, the POP method certainly has no problem with external validity. The model is specifically about this patient. Whatever we postulate about Ms Jones is 100% applicable to Ms Jones. After all our model is attempting to mirror a piece of Ms Jones' reality. It is an attempt to describe one area in that universe that is Ms Jones' world. In a sense when we are treating Ms Jones, we are engaged in the science of 'Jonesology'. When we are treating Ms Smith, we are practising 'Smithology'. The extent to which this last observation is a metaphor or literally true will be considered in more detail when the scientific status of POP model construction is considered in Chapter 10. Before then however it will help to further illustrate the usefulness as well as other features of POP models by considering several other cases.

Endnotes

1. American Psychiatric Association 2013, p. 222
2. The case for patient narratives being of value in medicine first through their transformation into causal models is described in the work of participants in Norway's CauseHealth project (Anjum *et al.* 2020).
3. This process of reasoning has some similarities to a process described in the field of hermeneutics as the 'hermeneutic circle'. Hermeneutics has a long history. It initially related to interpretation of texts, especially the exegesis of biblical texts. Since then, it has developed many branches with competing definitions. Most broadly it can be seen as the study of the methodology of interpretation in general (see Palmer, 1969; Phillips, 1996). The hermeneutic circle refers to a strategy of interpretation in which one moves from considerations of very specific facts to better understand generalizations and then uses generalizations to better understand the meaning of specific facts. By a process of moving back and forth between these, an increasingly rich and defensible understanding of a particular event emerges. A particularly influential branch of hermeneutics in the social sciences developed in the nineteenth century to define a methodology for

humanistic disciplines such as history posited to be distinct from the methodology of the natural sciences. While sympathetic to many aspects of hermeneutics, I think this particular methodological distinction is problematic. While viewing POP modelling as compatible with many hermeneutic strategies (see Sadler and Hulgus 1989), a key argument of this book is that the process of clinical reasoning based on model construction and revision is at one with the methodologies of the natural sciences, and not something opposed to it. This argument will be more fully developed in section 10.2.1.

4. Godfrey-Smith 2021
5. Scull 2022
6. Giere 1988, 1999
7. Weisberg 2014
8. Gluud 2006; Dunlop *et al.* 2012; Bhandari *et al.* 2004; Begg 1994
9. Lillie *et al.* 2011
10. Popper 1963.
11. Thornton 2023
12. Weisberg 2014
13. This type of reasoning is an instance of C. S. Peirce's concept of abduction as a mode of reasoning (see Burks 1946)
14. Cartwright 1999

Chapter 4
Changing goals and changing models

The case of Mr Z

4.1 Mr Z's initial presentation and early treatment

In the last chapter we noted that one of the strengths of pattern of propensities (POP) models is that their adequacy can be reassessed in light of the subsequent clinical experience with the patient. This allows refinement, elaboration, and revision of a model to better meet a patient's specific and unique goals, and to address changes in goals that may occur over time. The case of Mr Z highlights aspects of this process.

Mr Z first consulted me at age 43 at the request of his wife and their marital therapist for the evaluation and treatment of his depression. He described depressed feelings all his life. Both maternal and paternal relatives had significant histories of depression. He described his mother as especially critical and demanding. The household was dominated by her moods. Among the three children, he was her preferred target, with much emotional and occasional physical abuse. He was expected to tolerate her verbal tirades quietly and passively. If he challenged her even gently, he would be severely punished. He developed a style of quiet withdrawal, accompanied by an inner sense of emptiness, resentment, and anhedonia. Occasionally, however, he would explode with an angry tantrum usually over a relatively trivial matter. He put his energies into his studies, was an excellent student, and began a successful career as an environmental biologist.

In his mid-20s he became more depressed after a series of failed relationships with women, all of whom complained about his chronic anger and emotional unavailability. He had several years of individual psychotherapy that he found extremely helpful. He worked through many of the issues with his mother that had undercut his adult relationships with women. He specifically came to understand that his anticipation that all women would respond to him like his mother caused him to assume criticism from them whether present or not, and then to withdraw or lash out in anger. Their subsequent rejection of him simply confirmed his expectations. He gradually came to appreciate that his behaviour towards them provoked their rejection and constituted a self-fulfilling

How Psychiatrists Make Decisions. Douglas W. Heinrichs, Oxford University Press. © Oxford University Press (2025).
DOI: 10.1093/9780198960720.003.0004

prophesy. With this insight he developed some capacity to make himself emotionally vulnerable with women and engage in constructive communication.

During the time of his psychotherapy, a trial of tricyclic antidepressants was ended quickly due to intolerable side effects, but his mood improved and his anger diminished dramatically with psychotherapy alone. By age 30 he had married and by age 35 had two sons. About a year after the birth of the second child his wife developed a range of vague somatic symptoms including episodic muscle weakness and diffuse pain, anxiety, and insomnia. Extensive diagnostic assessments were inconclusive, with some sort of neurodegenerative or autoimmune disorder suspected but never confirmed. She began to receive increasing complaints about her work performance at the same academic institution where Mr Z worked. She also brought a complaint of sexual harassment against a supervisor that created considerable open conflict at work. The patient found all this very uncomfortable but supported his wife throughout. She ultimately left the job with a medical disability.

As her somatic complaints worsened, the patient found her increasingly demanding, unappreciative, and critical, although he dutifully took over more of the household and child-rearing responsibilities. He was increasingly sceptical as to the medical legitimacy of her somatic complaints as any clear diagnosis remained elusive and his wife seemed to find energy for an active social life with a group of friends apart from him. Nevertheless, he was devoted to her and committed to the marriage. Her complaints that he was more emotionally withdrawn, irritable, and angry led them into marital therapy. The marital therapist told me that she was struck with the level of his depression that interfered with his ability to participate in therapy, leading to the psychiatric consultation with me. He presented with depressed mood, impaired concentration, high irritability and anger, high anxiety, profound anhedonia, low energy, abulia, insomnia, and loss of appetite. He reported some suicidal ideation but he felt he would never act on it due to his sons. The patient agreed to a trial of antidepressant medication. After several selective serotonin reuptake inhibitors (SSRIs) made him feel fatigued and emotionally flat, he had a good response to venlafaxine at 300 mg daily. He described a marked reduction in depressed mood, better concentration and engagement at work and home, normalization of sleep and appetite, decreased anxiety and, most dramatically, the virtual elimination of his anger and irritability. He was extremely pleased and felt better than at any time in his life.

4.2 Incorporating multiple goals in a POP model

Mr Z presents, from one viewpoint, a rather textbook case of recurrent major depression to which recommendations based on EBM easily apply, leading to a recommendation of an antidepressant. I could have found it easy to view the goal of treatment as simply the alleviation of the patient's depression. However, from the vantage of Mr Z's goals for treatment, he was less focused on relief of his depressive symptoms than on saving his marriage. Worth noting is the potential connection between using an antidepressant and improving the odds of his marriage surviving. While this may seem clinically self-evident in this case, any evidentiary basis for the marriage-saving effects of antidepressants that would be recognized by EBM would be indirect at best. So how might I integrate these two goals for the patient into a treatment strategy? I would again suggest that my reasoning is best understood in terms of generating a POP model. The initial POP model I proposed is represented in Figure 4.1.

A condensed form of my initial presentation of this model to the patient is as follows:

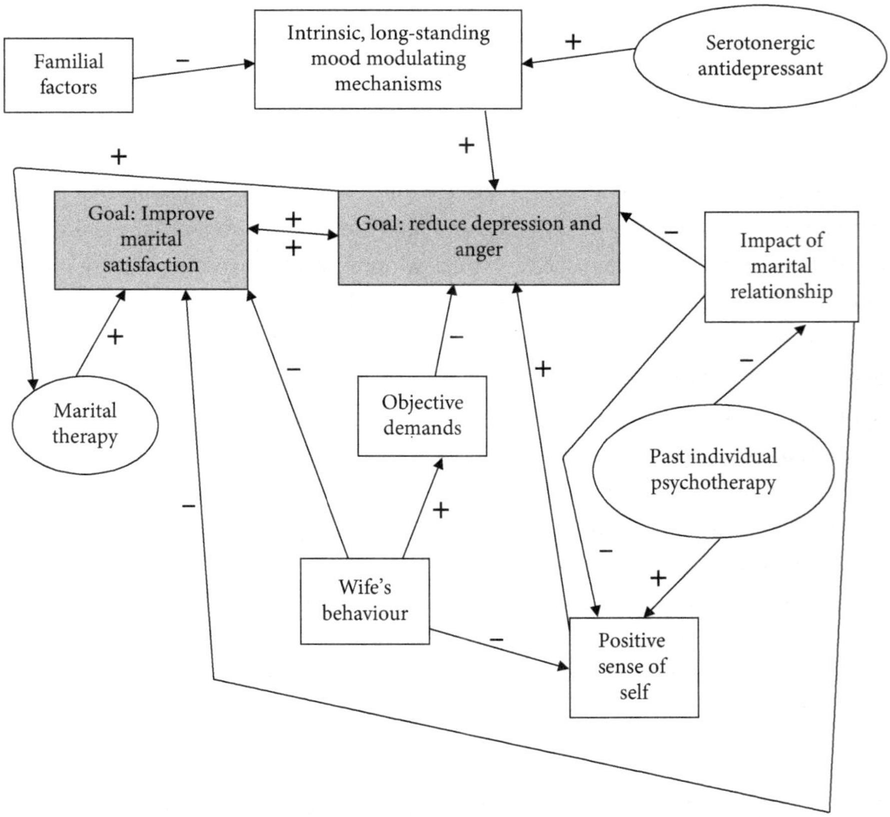

Figure 4.1 Initial shared model for Mr Z.

Reproduced with permission from Heinrichs, D. 2015. Model-based science and the ethics of ongoing treatment negotiations. In J. Z. Sadler, C. W. Van Staten, and K. W. M. Fulford (eds). *The Oxford Handbook of Psychiatric Ethics*. Oxford, UK: Oxford University Press.

Your strong familial predisposition to depression has been reactivated by the circumstances of your wife's illness and subsequent behaviour. Your wife's condition triggered it in two ways. Concretely it has burdened you with a high level of practical demands at home in addition to your demanding job. But it has also caused her to behave towards you in ways that remind you of your mother's treatment of you which was so depressogenic in the past. Whether your perception of her is fully accurate or distorted by your experiences with your mother is an open question that should be explored in marital therapy. In any case your depression is manifested in part by anger and emotional withdrawal that in turn makes the marital problems worse. I recommend an antidepressant to relieve some of your subjective distress and impaired functioning but also to reduce the anger and withdrawal, thereby improving the likelihood that marital therapy will be effective.

The situation of Mr Z is considerably more complicated than that of Ms K. There are really two interrelated goals that brought him to me. He was certainly interested in reducing his depression and anxiety symptoms and the distress they caused him. On the other hand, he was primarily concerned with improving the marital relationship for both himself and his wife. Hence the POP needed to include propensities and their interactions relevant to both goals. Furthermore, to further complicate matters as we shall see, the nature of these two goals and their interrelationship will morph considerably over the subsequent course of treatment.

4.3 A new model for new circumstances: Mr Z's later treatment

Over time, changing circumstances in a patient's life can require revisions of the model. That occurred for Mr Z.

Mr Z reported an ongoing excellent response to venlafaxine over the next five years. Yet he described continuing marital problems. Although not fighting, he felt his wife was progressively less interested in him and their sons, socializing with her friends often late into the evening. She was drinking more and he had reason to think she might be having an affair. She stopped marital therapy. He tried to talk to her but she was unresponsive. His younger son was beginning to show significant behavioural and emotional problems. He was feeling progressively more disappointed and emptier, yet showed no clinical signs of depression. Mr Z told me this with a sense of perplexity and frustration. He was especially struck by the almost total absence of visceral feelings or expressions of anger or irritability on his part, despite being very disapproving of his wife's behaviour. Initially he had ambivalent feelings about this. He was pleased to be spared the pain and disrupted functioning of a clinical depression as well as the anger, and felt the need to keep the home calm for the sake of his sons, especially the troubled younger one. However, he felt humiliated by his tolerance. He recalled similar feelings as a child when silently enduring mistreatment by his mother. Over time his ambivalence became a rather affectless outrage that became focused on the antidepressant: 'The drug stops my depression but I cannot get angry enough to stand up for myself. I feel gutless and emasculated and I cannot even get depressed about that. I want to stop my medication and let the chips fall where they may. I would rather be depressed and angry than like this'.

For a time, Mr Z was unwilling to consider any other course of action. However, he continued a discussion with me over several visits, culminating in my offering a revised model to the patient, the description of which can be condensed as follows:

> You have good reasons to be unsatisfied with the state of your marriage. Based on your past responses there is every reason to expect your current situation would have triggered irritability, anger, and a clinical depression if you were not on medication. It is most likely that it has not done so because of the action of venlafaxine. In the past we did not consider how this medication prevented your depression. We do not really know, but other patients have sometimes described a blunting of emotional experience and expression on some antidepressants, like venlafaxine, that primarily work by increasing the availability of serotonin at synapses. In fact, you felt unacceptably flat on several antidepressants that we tried initially, all of which were even more exclusively serotonergic. It is likely that the mechanism by which these drugs treat depression and anger involves, at least in some people, a blunting of their emotions. The results can be both good and bad. For a time, the positive consequences were overwhelmingly dominant for you. But now some of those feelings would be useful to mobilize you to deal with the situation and to avoid the shame and humiliation that comes from your passivity that reminds you of your childhood predicament. Of course, you can choose to stop venlafaxine. But you know the depression is likely to return, and do not lose sight of how crippling it was and how it too interfered with your ability to act effectively. I know you are feeling quite hostile to antidepressants right now but I would ask you to consider trying another type of antidepressant that relies less disproportionately on serotonergic effects. The hope would be to control the clinical depression without the emotional blunting. Of course, it may not work and you may become depressed. Even if it does work, you will have to deal with an array of feelings including considerable anger.

> I also think there is a high likelihood that you ultimately will leave your marriage with a difficult transition for you and your sons, but with the possibility of a sense of self without the feelings of inadequacy and shame, and the chance of finding a truly satisfying relationship in the future. Given the challenges ahead of you if you take this course, I would suggest ongoing individual supportive psychotherapy.

After further deliberation Mr Z, with considerable scepticism, decided to try a change of antidepressants and he started tranylcypromine, a monoamine oxidase inhibitor (MAOI).[1]

He experienced a significantly increased emotional intensity, felt more alive, and indeed became angrier. He entered psychotherapy. Within six months he filed for divorce. He was the primary emotional support for his sons, who experienced considerable emotional turmoil. The settlement process was adversarial and difficult. He experienced periods of anger, anxiety, disrupted sleep, and dysphoria but never became depressed to a degree that impaired his functioning. Now, several years later, he is generally happy and feels good about himself. His sons are doing well now, after an extended period of difficulty, especially for the younger boy. He still works to control his temper and irritability but is generally successful. He is dating casually but struggles with being able to trust in an intimate relationship. He remains in therapy and is working on this issue.

4.4 Coping with conflicting goals and a revised model

The subsequent events for Mr Z required a major revision of the POP model, one more sensitive to the complex facets of his case. Three related goals could now be distinguished: resolving his marital stalemate, establishing a positive sense of self, and avoiding incapacitating depression in the process. The postulated propensities and their interactions became significantly more complex as well. A diagrammatic representation of this new POP is shown in Figure 4.2.

The sequence of events in Mr Z's treatment illustrates the need for careful attention to the current form of the patient's distress and the changing ramifications for his goals and their treatment. The initial model focused in a rather straightforward way on the treatment of a classic major depression, albeit complicated by a series of issues in his family dynamics both past and present. With respect to the pharmacologic aspects of the treatment plan, the approach of EBM would work fairly well here. One could ask the question as to what are the best evidence-supported treatments for major depression

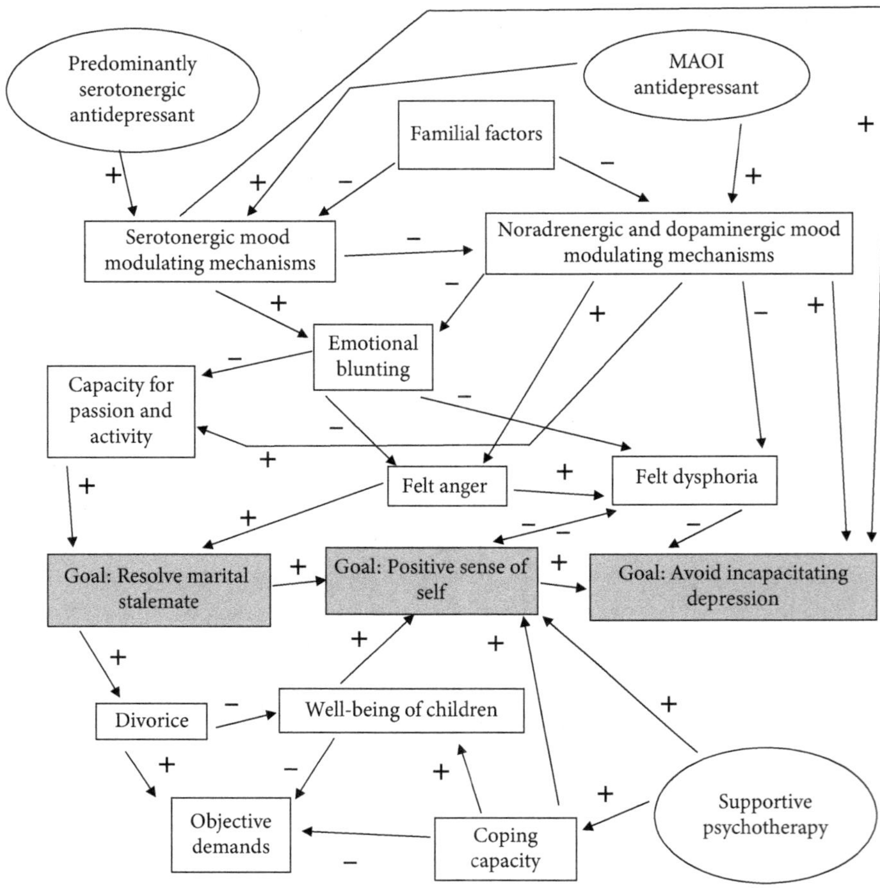

Figure 4.2 Revised model for Mr Z.

Reproduced with permission from Heinrichs, D. 2015. Model-based science and the ethics of ongoing treatment negotiations. In J. Z. Sadler, C. W. Van Staten, and K. W. M. Fulford (eds). *The Oxford Handbook of Psychiatric Ethics*. Oxford, UK: Oxford University Press

and it appeared reasonable to assume that results of those trials would be quite applicable to Mr Z. The range of symptoms of depression presented are quite typical as well. Indeed, after a series of trials of antidepressants he showed good response to venlafaxine with the reduction in his full range of depressive symptomatology.

However, as circumstances required major revision in his POP, the straightforward application of treatment options based on studies from traditional diagnostic categories became impossible. The constellation of symptoms characterizing depression in effect split apart. What the patient now wanted, and with understandable reasons, was to avoid the dysfunction that came with his prior depression but to have back some of the anger and dysphoria that accompanied that depression to motivate him into taking action that ultimately would make him feel better about himself. That was in fact accomplished. It is ironic that where both the patient and the psychiatrist would judge this second phase of treatment a success, had the patient been tracked with standard rating scales his condition would appear to be deteriorating. Furthermore, results of clinical trials are of little help here. One is unlikely to find trials assessing whether a given treatment will make some depressive symptoms worse but others better in a particular pattern. The circumstances are simply too unique to Mr Z to expect much help from 'answerable questions'. In contrast to a case like Ms K, the knowledge base one acquires in initial psychiatric training fails to provide on its own the range of imaginative options necessary to reformulate the model needed to guide Mr Z's subsequent treatment. I had to rely much more heavily on a wide range of subsequent experiences with other patients to generate other possibilities.

Nonetheless, one can still use our resources of more generalized knowledge in psychiatry to assess the evidentiary support for each hypothesized interaction between propensities in the model. One of the strengths of a POP model is that it allows the clinician to draw upon this body of knowledge even when the overall situation is highly unique to a given individual. We can ask questions such as: How strong is the evidence to support the prediction of a qualitatively unique response to an MAOI in this situation? What reason do we have to posit that supportive psychotherapy would help Mr Z weather the emotional storm of the ending of his marriage? This way of conceptualizing the use of general evidence to inform our thinking about an individual patient usefully contrasts the approaches of EBM and POP modelling. EBM attempts to provide a method to take generalized evidence from studies of groups of patients to directly answer the question 'What is the optimal treatment for Ms Jones?' In contrast, POP modelling uses generalized evidence as one approach to answering the question 'How strong is our basis for positing that propensity X has an impact on propensity Y as one link in our unique model for understanding Ms Jones?' In this sense POP modelling has a more fine-grained relationship to generalized evidence than EBM.

The evolution over time of Mr Z's situation and response to treatment raises an important ethical question. Given the propensity of strongly serotonergic antidepressants to cause a blunting of emotional experience and expression in some patients, should I have considered that possibility and its negative impact in advance, or at least have proactively looked for an unintended consequence of this sort before the patient reported it five years later? Even if I could not have anticipated the changes in his wife's behaviour over that time, should I have been actively questioning the patient about possible emotional blunting even during the time he said he was doing well? Was it a failure of my clinical imagination to not consider this possibility? Could I have spared Mr Z several years of unhappiness? However I or an outside observer assesses my culpability here, it is certainly true that since my treatment of Mr Z, I am considerably more attuned to this issue and proactively look for evidence of emotional blunting with serotonergic agents even in patients who report that all is well.

4.5 Levels of explanation: A theoretical conundrum

The case of Mr Z serves to highlight several other more theoretical characteristics of POP models. One is the issue of explanatory levels in models. This issue and its ramifications have been extensively discussed in fields such as the philosophy of science, whereas many clinicians have only been exposed to rather simplified formulations, such as Engel's mandate in the biopsychosocial model that the clinician should consider biological, psychological, and social factors in evaluating and treating a patient.[2] We are mostly left to imagine exactly how we are to integrate and weigh these various domains in reaching a coherent understanding of the patient and planning treatment.

A more developed understanding of the complex issues involved with the question of levels of explanation in science would assist the clinician in evaluating how one approaches the individual patient.[3] A great deal of theoretical writing has focused on the question of whether there is a fundamental or privileged level of explanation for psychiatric phenomena. More reductionistic thinkers argue that ultimately, as our knowledge grows, everything will be explainable in terms of fairly basic biological variables such as those of neurotransmitters, molecular biology, or neural networks.[4] Others argue that we will always need psychological, social, and cultural levels of explanation in psychiatry and have proposed strategies to integrate these explanatory levels—that is to relate one level of explanation to another and use them together coherently to explain psychiatric phenomena.[5]

Very little has been settled in this arena despite the extensive discussions. However, at the practical level of a POP model, it works perfectly well to include propensities that are described in terms of a wide range of levels. Indeed, one can argue that it is primarily in considering an individual case that integration of levels of explanation is feasible. As philosopher of science Nancy Cartwright observes, in most cases 'we must combine knowledge and technical know-how from a large number of different fields to produce a model that will agree well enough on the matters we are looking to predict, with the method of combination justified locally'.[6] Or in the words of the philosopher of biology Sandra Mitchell,

> . . . given the complexity and diversity of biological phenomena, simple models are more likely to capture partial causes. In this situation, individually more complete models may be developed if there are means for replacing the false assumptions in the model by more realistic descriptions . . . More adequate description will require integrating diverse theories of all the relevant features that uniquely characterize a given phenomenon . . . Hence integration of diverse approaches is likely to be one hallmark of better science.[7]

The basic thrust of these arguments is that however one views the feasibility and desirability of reducing one level of explanation to another at the general or global theoretical level, our best chance of capturing as fully as possible the complexity of a unique situation with a functional model requires us to be free to combine and move between different levels of explanation in whatever way is most useful to achieve our goal. Genetic and biological factors can interact side-by-side with psychological and social considerations in an unproblematic way. It becomes a pragmatic issue as to what levels best describe the propensities in question. In the initial POP for Mr Z, familial factors were thought to contribute to difficulties by some disruption of an intrinsic biological mood regulating mechanism leading to his propensity to chronic depression. Furthermore, it was unproblematic to also suggest that early life experiences and current relationship issues could interact with those biological factors. Treatments were offered, such as antidepressant medications, without much need to consider the specific mechanisms by which those antidepressants influence the underlying biology of the patient. However, with the more complicated circumstances leading to the revised POP model, it became useful to hypothesize a more complex model for the action of antidepressants that required speculating about several aspects of mechanisms of action in order to predict differential response between an MAO inhibitor and a predominately serotonergic antidepressant. This required recasting this piece of the model in terms of what would be viewed as a more basic level of explanation involving neurotransmitter behaviour. In keeping with the pragmatic nature of POPs, the choice of levels on which propensities and their interactions are described depend upon the practical goals of the treatment. This raises the interesting theoretical possibility that there are no intrinsically privileged or fundamental levels of explanation and that the attempt to integrate entire levels to one another cannot be done in the abstract. Perhaps relating various levels to one another only make sense in the context of the consideration of a concrete individual instance, such as Mr Z.

4.6 Two concepts of diagnosis

If the POP approach to clinical reasoning focuses on a unique individual modelling of the given patient's circumstance as opposed to regarding him as an instance of a larger diagnostic category, what then is the status of a diagnosis in psychiatry? What are we claiming in assigning a diagnosis to a patient and what purpose do we expect it to serve? I would suggest that while a patient cannot be reduced to an instance of a particular diagnostic group, a diagnosis can in some situations be useful in the construction of the POP itself. Ultimately a diagnosis will only be meaningful if we posit that individuals who have that diagnosis share some cluster of propensities that contribute to their distress. The fallacy is seeing the diagnosis as describing the entire relevant set of propensities in any given case. In assigning a diagnosis in the context of POP modelling, the clinician is postulating that some hypothesized cluster of propensities implied by that diagnosis, however ill-defined, is operating in this patient and form some piece of the larger POP of the patient.

In the case of Ms K, it was hypothesized that she shared some cluster of biologic propensities on a familial basis with individuals who met criteria for generalized anxiety disorder. Even though the specific details of those propensities at a biological level are poorly understood, the diagnosis itself can serve as a place marker for them and figure into the POP model. In the case of Mr Z, the criteria for major depression were clearly met. Hence based upon what knowledge we do have about individuals who meet this diagnosis it could be proposed that he had certain propensities, possibly on a genetic basis, that form a piece of his larger POP. In the revised and elaborated POP for Mr Z, the speculations about that depressive propensity had to be expanded, again drawing from themes in the depression literature[8] which suggest a role for multiple neurotransmitters as well as providing a hypothesis about potential differential responses to different types of antidepressants.

At first glance this may seem a curious use of diagnoses given that the DSM (*Diagnostic and Statistical Manual of Mental Disorders*) officially argues that they are atheoretical descriptive entities that posit no clear underlying mechanism.[9] Indeed, at this point that is how the DSM is constructed. However, even the authors of the DSM see this as an interim arrangement and anticipate that as our knowledge grows underlying explanatory factors will emerge. The diagnosis is then essentially a promise of future understanding and becomes a place marker in an individual patient's POP for certain poorly characterized propensities that the patient is thought to share with some group of patients. These in turn interact with the larger array of propensities that provides the unique model for this individual patient.

It is crucial to remember that even if the explanatory understanding place-marked by a DSM diagnosis were to be fully elaborated by future research, this would fill in just one piece of the unique pattern of propensities that model a given patient's situation. Thus for Mr Z, I can confidently propose that we will never discover a shared mechanism for a diagnosis of 'obsessional men who are depressed and lonely and angry at their disengaged spouses and whose depression but not anger responds to MAOIs'. We can reduce potential confusion here by separating two senses of the term diagnosis as distinguished by Sadler[10]: diagnosis as the categorizing of a morbid condition in contrast to diagnosis as a clinician's complex understanding of the unique situation of a specific patient. The former is the DSM concept of diagnosis while the POP model is the latter.

This further implies that how important the diagnosis in the DSM sense is for a particular patient will vary substantially depending upon the goal of treatment. Consider for instance the case of an individual with chronic schizophrenia who is well stabilized on antipsychotic medication and functioning fairly well who suffers the unexpected loss of a parent with whom he is quite close. In this case the patient's diagnosis may have relatively little impact on his immediate goal which is helping to resolve his grief. Other than the prudent need to be alert to any evidence of a decompensation being triggered by the loss, his diagnosis may play very little role in the POP model directed towards resolving his grief.

4.7 Pattern recognition

As a final consideration of the case of Mr Z, I again want to draw attention to the role of the recognition of patterns of occurrences in developing the model. Mr Z's life from early on until the present has been dominated by recurrences of depressive manifestations. Also, certain sorts of triggers for his worst depressions have recurred throughout his life, initially the way he was treated by his mother followed by the recurrence of that treatment in varying degrees in other relationships with women and most prominently with his wife. Put another way, one would expect that a propensity of significant strength would manifest itself by characteristic affective or behavioural reactions in response to multiple similar circumstances that are likely over a patient's life. More frequent recurrences of such a pattern strengthens confidence in the existence of the proposed propensity. Thus, for Mr Z, as in the case of Ms K, our confidence in the POP model is enhanced by both the presence of recurrent patterns as well as the extent to which predictions based on the model prove to be true.

Endnotes

1. For the non-psychiatrist reader some background information may be helpful here. The primary mode of action of most current antidepressants is thought to be increasing the availability at the synapse of one or several neurotransmitters termed monoamines—primarily serotonin, norepinephrine, and dopamine. This increased availability induces a cascade of events that ultimately results in the reduction of depressive symptoms (Racagni and Popoli 2008). The usual first-line antidepressants are a family known as selective serotonin reuptake inhibitors (SSRIs). Their popularity largely results from their high safety profile. The selective increase in availability of serotonin appears to be very useful in treating anxiety symptoms as well as depressive ones. Their weakness appears to be in consistently addressing lack of pleasure (anhedonia), low energy, low motivation, and apathy. In some cases, SSRIs actually appear to worsen these dimensions with an increase in emotional blunting and indifference (Sansone and Sansone 2010). Mr Z showed this worsening on SSRIs. The antidepressant venlafaxine was chosen because it enhances the availability of norepinephrine as well as serotonin, although its impact on the former is significantly less than on the latter. Enhancing norepinephrine and dopamine are thought to address these dimensions of the depressive syndrome more effectively than enhancing serotonin alone, and to neutralize the serotonin-induced worsening of these features (Nutt, 2008). Indeed, Mr Z felt less blunted with venlafaxine and was quite satisfied for several years with this treatment. When it subsequently became apparent, as Mr Z confronted his marital crisis, that emotional indifference was still a significant problem, the switch was made to an antidepressant thought to have a stronger impact on norepinephrine and dopamine. Monoamine oxidase inhibitors (MAOIs) work by deactivating an enzyme that breaks down all three monoamines, thereby inducing increased availability for all of them. They are quite effective antidepressants, but are less popular due to the risk of severe toxic interactions with several other medications and some foods.
2. Engel 1980
3. Anyone thinking about the nature of science quickly observes that each scientific discipline has its own theoretical entities and theories of how those entities interact to explain the phenomena studied by the discipline. But how can we combine these theories to achieve a comprehensive scientific view of the world? And is this in fact possible? The rapid success of modern science in explaining an increasing range of phenomena led early philosophers of science to assume that such a unification of scientific theories into a grand scientific explanation was possible if not inevitable as science grew. The contemporary

philosopher of science Nancy Cartwright (1999) refers to this attitude as 'scientific fundamentalism', seeing it as more like a religious belief and hope than a position based on evidence. Attempts at grand integration theories typically start with the observation that there appears to be a natural hierarchy among sciences. Physics appears to be the most basic and universal discipline, followed by chemistry, biology, psychology, and sociology. The challenge then is to show how theories at the higher level can be reduced to those at the lower level. For instance, the laws of chemistry should ultimately be explainable in terms of or 'reduced to' the laws of physics. If we cannot yet do this, it simply reflects our incomplete knowledge of the laws of physics—a gap that will be filled in eventually as the science grows. By the same logic, biology should reduce to chemistry, psychology to biology, and sociology to psychology. Furthermore, each discipline that constitutes a level of explanation has within itself multiple sublevels of explanation that can also be arranged in a hierarchy. Thus, with enough knowledge, classical physics should be completely explainable in terms of quantum physics, neurochemistry in terms of molecular biology, and so forth. More contemporary philosophers of science, especially those interested in higher level disciplines like biology and psychology, have argued convincingly that there are good reasons to believe that this kind of single unified theory based on reduction is impossible. They are then left with the task of explaining how the various levels of explanation relate to one another and how they can be used together in an integrated and coherent way. For useful and accessible discussions of these complex issues for the interested clinician, see Cartwright 1999, Mitchell 2003; Schaffner 1993

4. Churchland 1982; Gold 2009
5. Karlsson and Kamppinen 1995; Kendler 2005; Schaffner 1993, 2016; Murphy 2008; Owens 1989
6. Cartwright 1999, p. 10
7. Mitchell 2003, p. 190
8. Sansone and Sansone 2010; Racagni and Popoli 2008; Nutt 2008
9. American Psychiatric Association, 2013, p. 20
10. Laney and Sadler 2021

Chapter 5
Models and patient values

The case of Mr D

5.1 Prioritizing patient values and preferences with pattern of propensities (POP) models

Psychiatrists and physicians in general are frequently admonished to take patients' values and preferences very seriously in planning and implementing treatment. We have seen in section 2.4.3.4 how evidence-based medicine (EBM) tries to incorporate this by generating some numerical value for this dimension that can be plugged into a formula, the results of which are meant to help the patient in decision-making. I have argued that this kind of pseudo-quantification makes little intuitive sense, and hence is likely to offer little benefit for the patient. Furthermore, multiple preferences and values come to play a role in a patient's decisions about treatment options. This cannot be meaningfully captured by a single numerical value. In the examples presented thus far of clinical reasoning based on POP model construction, this dimension of concern for patient values and preferences is incorporated in several ways. First a meaningful model can only be constructed in reference to a specific goal or goals. A working relationship with the patient can only be based upon goals being shared between doctor and patient. The construction of a POP model encourages the clinician to explicitly discuss the goal or goals of treatment with the patient at the very beginning of the treatment relationship. In the case of Mr Z, the goals needed to be revised over time considering the subsequent developments in his life. Once goals are agreed upon, a POP model encourages the psychiatrist to elaborate a hypothesized pattern of interacting propensities thought to explain the development of the patient's distress and thereby suggesting points of intervention to relieve that distress. By sharing this model with the patient, the rationale for proposed treatments becomes clear, and their acceptability to the patient can be fully discussed. In both the cases of Ms K and Mr Z, this process led to some modification of the treatment plan given the patient's preferences and priorities.

How Psychiatrists Make Decisions. Douglas W. Heinrichs, Oxford University Press. © Oxford University Press (2025).
DOI: 10.1093/9780198960720.003.0005

5.2 Mr D's initial presentation

Sometimes, however, the distance between the psychiatrist's conceptualization of the patient's issues and the patient's own assessment can be quite divergent and requires a more radical degree of rethinking and renegotiation on one or both parts. This is illustrated in the case of Mr D.

Mr D entered treatment with me at age 40, shortly after discharge from the hospital. A bright man diagnosed as suffering from paranoid schizophrenia, he was able to hold a highly responsible technical job despite being very socially isolated. When not at work he spent his time decoding comic strips for communications from aliens threatening to destroy the earth, hoping to learn how to stop them. Usually, he was able to approach this task with detached intellectual curiosity. Periodically however he became overwhelmed at the potential threat from the aliens. He then began preaching in the streets and telling people to seek shelter, or he was taken to the emergency room in a frantic, agitated state. Several weeks of hospitalization and treatment with an antipsychotic restored his calm state and allowed him to re-establish his detached interest in his delusions. However, shortly thereafter he would invariably stop medication because he began to feel too mentally dulled and flat, and increasingly detached from his delusional interests. The pattern would inevitably repeat in 6 to 12 months, culminating in another hospitalization.

Mr D had worked with several outpatient psychiatrists but always terminated treatment when they argued that his beliefs about aliens were untrue and insisted that he take continuous maintenance antipsychotic medication. Initially Mr D wanted no medication since the aliens were real in his mind and the medication never took his belief in them away, but it did disrupt their communications with him with fewer messages from them for him to decode. He went on to explain that he had no real life outside of his work, but his interactions with the aliens and his mission to decode their messages provided him with a sense of purpose as well as intellectual stimulation. When medication made it impossible for him to engage in this activity, his life seemed empty and boring. He wanted a psychiatrist who would be willing to treat him without medication. He also wanted a person he could talk to about the aliens, which he found both stimulating and comforting. He realized that to do so with coworkers or strangers would be perceived as 'crazy'. Nonetheless, he felt a great pressure to be able to discuss these issues with someone. He assumed the psychiatrist would see the aliens as symptoms of mental illness, until perhaps persuaded otherwise, but would have to listen anyway since 'after all, I am paying you to listen'.

5.3 Finding an honestly shared model is critical

Until Mr D voiced his clear treatment preferences and the reason for them, I had begun formulating in my own mind a POP model with the treatment of psychotic symptoms as an important goal, along with maintaining the patient's functional capacity. Such a model could draw heavily on EBM sources for the use of antipsychotic medication in schizophrenia. Such a model is illustrated in Figure 5.1.

Once the patient's preferences were made clear, that model was untenable as a shared view. After all it was the refusal by past psychiatrists to give up that model that led him to terminate treatment with them. On the other hand, the clear recurrent pattern of hospitalizations made it overwhelmingly likely that

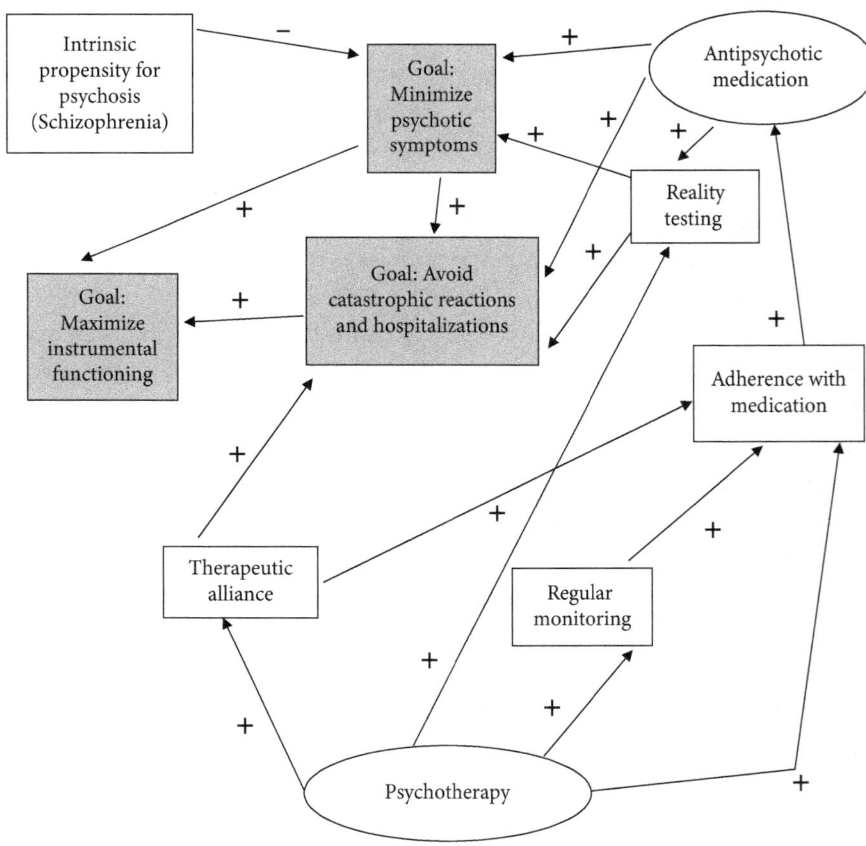

Figure 5.1 Psychiatrist's initial model for Mr D.

Reproduced with permission from Heinrichs, D. 2015. Model-based science and the ethics of ongoing treatment negotiations. In J.Z. Sadler, C. W. Van Staten, and K. W. M. Fulford (eds). *The Oxford Handbook of Psychiatric Ethics*. Oxford, UK: Oxford University Press

without medication Mr D would continue his pattern of returning to the hospital in highly distressed, agitated, and psychotic states, with the associated disruption to his life and functioning. Therefore, I proposed to the patient a model that can be condensed as follows:.

> Your experience of the aliens can be viewed as either caused by real external realities or as products of your own mind. We may never agree on which is true. We can agree that sometimes those experiences overwhelm you with a dread and desperation that is intolerable. Furthermore, we can agree that a relatively short course of antipsychotic medication and the opportunity to discuss your experiences with others restores calm in the sense that you can cope with these experiences and go on with your life. I have seen this with many persons with similar sorts of experiences. After treatment, most individuals feel calmer and stronger. Some of them come to believe that their experiences had in fact been products of their own minds while others remain firmly convinced that they come from external entities. Perhaps it does not matter that much, so long as you feel calmer and can function well. Hence, I propose that we meet regularly to talk about your experiences. If you start to get too upset by them, you can take medication briefly until you feel calm and then stop. Our job is to intervene before matters escalate to a point where hospitalization is necessary.

This model is illustrated in Figure 5.2.

Mr D found this arrangement agreeable. Initially he held out the hope that he might ultimately be able to convince me as to the reality of his beliefs about the aliens, but said that the arrangement would meet his needs regardless so long as I did not try to convince him they were false. My approach was to ask him detailed questions about how he felt the world of the aliens operated and how he interacted with it. Over time, he became less and less concerned with trying to convince me about the reality of his beliefs. He seemed to enjoy the role of educating me about this world that was so important in his life. He met with me at least monthly for a full psychotherapy session, taking low doses of antipsychotics intermittently, for the next 24 years without further psychiatric hospitalizations until his death from cancer.

5.4 Some keys to successful model negotiation

In the case of Mr D, my initial model had to be revised significantly, both in terms of goals and treatment interventions, to find a position on which we could concur. However, such a position was found and proved successful

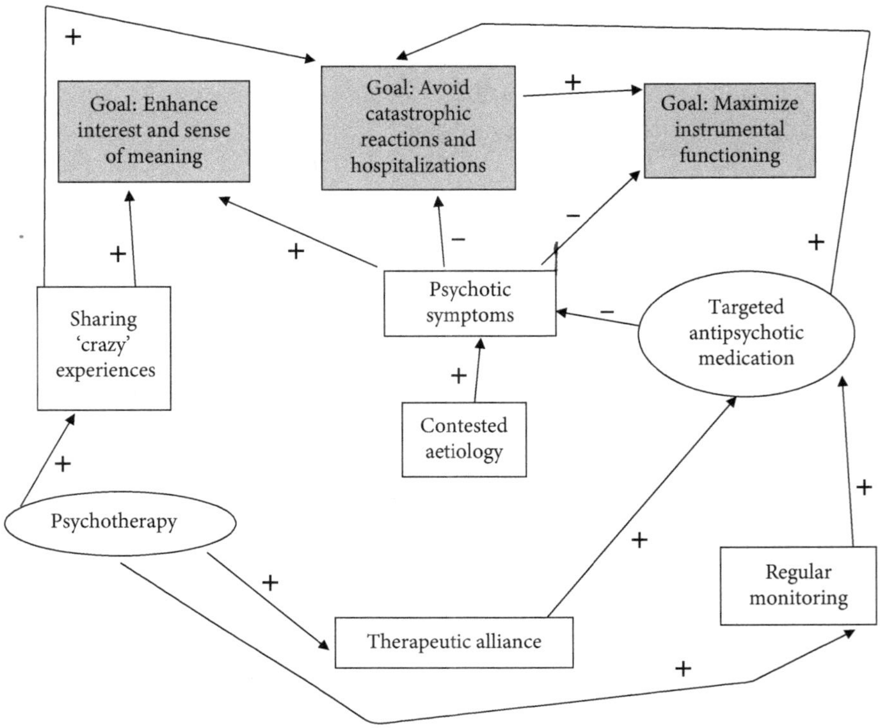

Figure 5.2 Shared model for Mr D.

Reproduced with permission from Heinrichs, D. 2015. Model-based science and the ethics of ongoing treatment negotiations. In J. Z. Sadler, C. W. Van Staten, and K. W. M. Fulford (eds). *The Oxford Handbook of Psychiatric Ethics*. Oxford, UK: Oxford University Press

for many years—indeed, for the rest of his life. Over time, I came to see the shared model as less of a tenuous compromise and more of a true collaborative understanding. I also developed a deep appreciation of one of Mr D's unique strengths that made this succeed. While he lacked insight in the traditional sense as to the unreality of his delusions, he consistently maintained insight as to how his beliefs would be perceived by others. He appreciated the need to refrain from discussing these beliefs with coworkers and other people with whom he came into contact. Indeed, he used the psychotherapy setting as a place where these views could be safely discussed without adverse consequences to his life. This was key to the success of our collaboration.

Another key to our success was his ability to live with the tension of our different beliefs. While I did not try to persuade him that his beliefs were false, I never pretended to believe them myself and he clearly understood that I did not. What became an increasingly explicit modus operandi for us over time was that when our two viewpoints both pointed to the same course of action, we should take it, even if our respective understandings as to why this was the thing to do differed. Thus, when he became emotionally overwhelmed by the messages, he briefly used antipsychotic medication. From his perspective it strengthened him to tolerate the ominous implications of the messages. From mine, they reduced the emotional intensity of his delusions. Whoever was correct, medication was indicated. This approach also saw us through what was one of the biggest crises in our years of working together.

About five years before his death, he revealed to me that he had a long-standing practice of going to the racetrack and betting on horses. He spent a great deal of time studying the horses, and over the years he had shown a modest net profit from this activity. He told me that in the preceding six months the aliens had begun to send him messages as to how he should bet on the horses. They assured him that they were more knowledgeable than he was and that he would ultimately make considerably more money. He began following their advice but was losing more and more money by doing so. I questioned him as to the wisdom of his continuing to follow their advice, but he continued to do so and dismissed my concerns as reflecting my lack of belief in the reality of the aliens. Several months later, I received a frantic call from Mr D. He had lost a large amount of money following the advice of the aliens. He felt deep despair and reported significant suicidal ideation. We worked through this crisis by appealing to the idea that if the same course of action could be justified from our respective viewpoints, it should be followed. If the aliens did not exist, then clearly it made no sense to follow their advice. If they did exist, however, they either grossly overestimated their knowledge of the horses or were intentionally and malevolently misleading him. Either way, he should stop following their advice. He was able to accept this and stopped losing money. Fortunately, he had not yet put himself into a financial situation from which he could not recover and his despair and suicidal ideation resolved.

5.5 An example of negotiated pruning

Mr D represents a particularly obvious case of negotiated pruning. The option of maintenance antipsychotic medication was taken off the table as unacceptable to the patient. While I originally tended to see this as simply reflecting his delusional beliefs as to the reality of the aliens, over time I came to accept the reasonableness of his position. He was a man who was quite lonely and fundamentally bored by his day-to-day life. He enjoyed using his mind and his intellectual bent was as a puzzle solver. His interaction with the aliens provided both intellectual stimulation and a sense of meaningful interaction with others as they communicated with him.

Sometimes, of course, no shared model can be found that is acceptable to both doctor and patient. Even in these cases, however, some limited benefits result from making the proposed models explicit. By giving the patient the sense that the clinician is not rejecting the patient's preferences in a knee-jerk manner but is making a thoughtful attempt to conceptualize the patient's situation, a foundation of mutual respect may open the door to future collaboration even if no shared course of action can be negotiated now. It's especially helpful for clinicians, in explaining why they cannot in good conscience comply with a patient's request, to distinguish between being unwilling to accept the patient's goal as a legitimate one (e.g. a college student who wants stimulants to study all night before a final exam) and accepting the goal as reasonable but judging that the requested intervention will fail to achieve it or risk unacceptable consequences (e.g. an actively drinking, depressed alcoholic outpatient rejects an antidepressant and wants benzodiazepines to reduce anxiety in the hope of then not needing to drink).

The case of Mr D also illustrates clearly how a shift in the goal of treatment may radically change the relevance of available data from clinical trials or other research. I very briefly considered an initial POP model in terms of eliminating or at least reducing schizophrenic symptoms, in this case specifically delusions. A large body of relevant studies considers the treatments, especially maintenance antipsychotics, that may offer some benefit here. With the revised shared goal, very little guidance from the scientific literature exists. The answerable question becomes: what is the best treatment for schizophrenic patients that will not take their symptoms away but will prevent them from becoming episodically overwhelmed and needing hospitalization? Little in the way of clinical trial data addresses that. Interestingly, of course, the large and growing body of evidence as to the potential negative effects of long-term antipsychotic use actually provides a good case in support of Mr D's preference for intermittent and targeted use of antipsychotic medication, even if this did not play a role in his own thinking.[1]

5.6 Ethical exercise of expertise

Mr D highlights in a particularly striking way a set of ethical issues that are present in all our encounters with patients. An intrinsic tension always exists when a patient sees a doctor who is defined as an expert in the condition for which the patient seeks help. As an expert, it is easy for doctors to assume they know what is best, encouraging what has been the long-standing tradition of parentalism in medicine. In recent years there has been a growing tendency to make the patient's autonomy primary and see patients as possessing a form of expertise based on their experience of living with their illness—'expertise by experience'.[2] Patients are seen as having a right to make the ultimate decisions, within broad limits, about matters that will directly and profoundly affect their own life and well-being. While to an increasing degree doctors acknowledge this principle of patient autonomy, clinicians may be tempted to think, given their role as experts, that disagreeing patients are in error. We may feel that the patient has 'a right to be wrong', but still believe our job is to persuade the patient, if possible, to accept our notion of what is best rather than strive for true shared decision-making.[3] As medicine, including psychiatry, has come to see itself as a more scientific activity, it is easy to buy into this notion of there being an objective truth of the matter. After all, isn't that what science seeks to achieve? In Chapter 10, I will argue that this represents a serious misunderstanding of what science is and does. At this point it suffices to say that, especially in a field like psychiatry, a considerable range of possible and reasonable options may be presented by us to our patients and to us by them. This becomes the starting point for truly shared decision-making.

One of the strengths of the POP model is that it starts with articulating the specific goals of the patient. Only in relationship to those goals can a model be proposed and a treatment plan based upon that model be judged as reasonable or applicable. Mr D's goal is not the one that a psychiatrist assumes to be operative with most schizophrenic patients; however, many more patients have goals of this sort than psychiatrists may appreciate. After all, Mr D was an exceptionally bright and articulate man. We might speculate how much of the notoriously high rate of nonadherence in psychotic patients comes from the failure of psychiatrists to develop a shared and explicit articulation of the actual goals of the patient.[4] While sometimes those goals will be such that the psychiatrist cannot embrace them, perhaps more often than we realize some unique but reasonable goals can be found, upon which the model and a collaborative treatment plan can be built.

Endnotes

1. Harrow *et al.*, 2021
2. Slade, 2017
3. Legare *et al.*, 2018
4. Scholl *et al.*, 2018; Stomski and Morrison, 2017

Chapter 6
Shifting and conflicting models

The case of Ms M

6.1 Ms M's initial presentation

The following case further illustrates the ability of a pattern of propensities (POP) modelling approach to articulate the clinical reasoning process and help to clarify a complex case that over time involved two psychiatrists with fundamentally different understandings of the patient's condition and optimal treatment.

Ms M began treatment with me at the age of 34. She was an extremely successful professional in the IT world. She worked very hard and progressed rapidly with many promotions and considerable financial success. She was seen as a rising star in her industry. Her personal life was considerably more chaotic. She described a series of numerous failed relationships going back to late high school. In each case she became deeply emotionally involved with a man quickly but over a matter of months would become increasingly arrogant, demanding, and self-centred in the relationship. Ultimately, the man would leave her after which time she became deeply depressed and at times suicidal. She came into treatment with me after the end of such a relationship. She was depressed and hopeless. There was some fleeting suicidal ideation without active intent. She stated that her work performance was falling and that at times she was not attending to her daily needs such as showering and washing her hair regularly. When not at work she stayed in her apartment in sweat clothes and did nothing except watch meaningless television and eat junk food. She stated that she felt emotionally depleted without a man in her life, could not take any pleasure in her accomplishments, and lacked any sense of self-worth. In the past these episodes following a break-up could last weeks to months. She then would typically become gradually more mobilized, able to get out socially, and reasonably soon would enter another relationship.

As soon as she was in another relationship the depression alleviated immediately and completely. She described that over time in a relationship her low self-esteem transformed into a rather inflated sense of self. She then would feel that she deserved more

How Psychiatrists Make Decisions. Douglas W. Heinrichs, Oxford University Press. © Oxford University Press (2025).
DOI: 10.1093/9780198960720.003.0006

and more from the relationship and that the man was lucky to have her, if not indeed unworthy of her. She became more stridently demanding in the relationship. This process usually evolved over several months, culminating in the ending of the relationship when she immediately plunged into another depression. When she was not depressed, she described herself as being extremely driven at work and being known as a remarkably high achiever. She did have occasional interpersonal problems in the workplace saying that at times other people found her too arrogant, opinionated, and impatient. Shortly after beginning treatment the patient improved enough to enter another relationship and I could see firsthand the transformation the patient had described. She became haughty, arrogant, and rather intensely driven. There was no rapid speech or prominent pressured thinking. She slept 6 to 7 hours a night which she described as her normal sleep pattern when not depressed. There was no evidence of loose associations, tangential thinking, or psychotic processes. Other than her arrogance and haughtiness there was no other gross impairment in judgement. Other than in the context of the beginning or ending of a relationship there were no episodes of significant mood swings in the patient's life.

Ms M grew up as the only child of what she described as a very passive and dependent mother and an extremely driven, highly successful scientist father. Father was also very handsome and charismatic and was repeatedly unfaithful to the mother. When the patient was eleven years old her father left her mother and moved in with his girlfriend, but maintained regular contact with the patient. Ms M had a reasonable relationship with her mother and felt sorry for her. However, she devalued her mother as not good enough to keep the father. Her relationship with her father was highly ambivalent. She was totally taken in by his charm and charisma and very much wanted to be adored and valued by him. On the other hand, she felt that no matter what she achieved, it would never be seen as good enough, as he could be highly critical of her despite her professional success. As she described it, the relationship appeared to have a considerably flirtatious quality in the way the father dealt with her although there was no sexual abuse. Other than the father's haughty arrogant nature, there was no other history of mood disorder in the family.

6.2 The preliminary model

I noted several features in Ms M's story that were suggestive of a bipolar disorder. However, considering the family dynamics and the way in which her shift in symptoms was so intimately tied to her relationship status, I felt an alternative model was more likely to characterize her problem.[1] In condensed form, I presented this preliminary model to the patient as follows:

> While your quite dramatic shifts in mood could be suggestive of some sort of bipolar mood disorder, I think an alternative understanding is more likely to be useful. From your description it appears that you greatly admired your father but could never feel secure in the knowledge that he valued you. You realized that your mother certainly could not compete successfully with the other women in his life. It is possible that your own extreme drive and ambition is an attempt to win his approval. That approval became tremendously important for your sense of self-worth. I think it is likely that this pattern plays out in your adult relationships. When you have a man in your life, a magnified sense of self-worth leads to a certain arrogant, condescending, and demanding quality that ultimately alienates the man such that he ends the relationship. When that happens, your self-esteem drops from the very top of the scale to the bottom. This leads to a deep depression during which you essentially shut down and can barely function. You gradually improve enough to be able to find another relationship and the cycle repeats. If this is a correct understanding of what is happening, I think there are two elements that should be part of your treatment. Far and away the most important intervention would be intensive psychotherapy to explore the long-standing and complicated relationship issues with your father and to be able to relate to men in your current life in a different way to allow a sustainable relationship. In the course of this work, I would hope that you would be able to develop a more realistic and stable view of your own self-worth. You may come not to see yourself quite as grandly as you do when in a relationship but certainly not feel as worthless as you do now. With the stabilization of your self-image, it is hoped that your moods stabilize as a result. Secondly, there may be a limited role for antidepressant medication to help bring you out more quickly from a depressed episode and perhaps to make it less likely that you will fall as deeply into one if another relationship ends.

A diagrammatic representation of this POP, including the recommended treatments interventions, is presented in Figure 6.1.

6.3 Testing the model with predictions

This model is directed towards two goals as put forward by the patient: the stabilizing of her mood and associated distress, and acquiring the capacity to sustain a healthy relationship with a man. Those goals clearly and directly interact with one another. By articulating the model, the points are evident regarding where the hypothesized explanation is testable. As always, whether the proposed treatments bring about the desired goals is the ultimate assessment of the model's adequacy. However, other occurrences may reflect upon

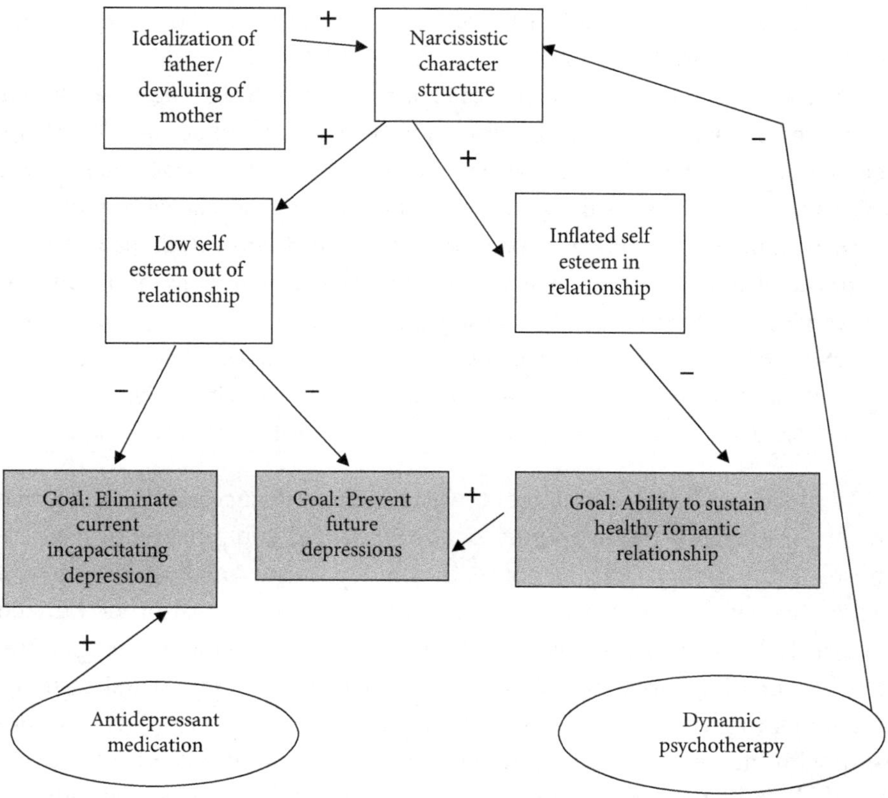

Figure 6.1 Initial model for Ms M.

the accuracy of the model. If during treatment, the patient has a major episode unrelated to her relationship status it would argue against the adequacy of this model. If the introduction of an antidepressant initiated a clear hypomanic episode or led to a pattern of increasingly frequent mood swings beyond her baseline, it would suggest that a disposition to a bipolar disorder should play an important role in the model. Finally, patterns of interacting with the treating psychiatrist could demonstrate transference issues either compatible or incompatible with this model.

Initially, Ms M agreed to the treatment recommendations. She was begun on a selective serotonin reuptake inhibitor antidepressant (SSRI) and in fact her presenting depression resolved faster than what was typical in the past. There was no evidence of any hypomanic or manic symptomatology after the introduction of the antidepressant. Several months later the patient became involved in another relationship, and shortly after that began to demonstrate the haughty, arrogant attitude that she had described at her initial assessment.

The psychotherapy component proved to be much more challenging. When she was depressed, she appeared to engage diligently in therapy work and seemed interested in pleasing me. Over time however as her depression lifted, and especially after she was in another relationship, she became more arrogant, demeaning, and dismissive of me. For periods the work would become very unproductive and inevitably the patient would begin cancelling and missing appointments and ultimately interrupting the therapy. She did continue the antidepressant medication however throughout. Her pattern of relationships continued as before. Each time a relationship ended she became depressed and returned to therapy only to pull away again when in a new relationship. After about 18 months, and while in a relationship, she terminated treatment with me abruptly.

While this phase of treatment was clearly not successful, much that emerged during it provided empirical support for components of the proposed POP model. The response to antidepressant medication was not suggestive of an exacerbation of an underlying bipolar disorder but did appear to help her depressive symptoms. There were no mood swings unrelated to relationship status. Finally, the patient in therapy related to me, a man, in a way very parallel to the way the model characterized her relationship with men in general. Where the model failed was in my estimate of the strength of the transference toward me and my inability to effectively address it before she terminated therapy even though I expected it to emerge. I initially felt the odds of her getting past this issue with a male therapist was better than the chances of her valuing and making use of the input of a female therapist, considering her prominent devaluation of her mother. What I failed to appreciate in my initial model was the extent and speed of her devaluation of me once she was in another relationship, such that she no longer gave serious consideration to anything I would interpret to her. Thus, this aspect of the model would subsequently need revision.

6.4 An alternative model

Shortly after terminating treatment with me, as another relationship ended and another depression ensued, Ms M consulted a nationally renowned expert in bipolar disorder and its treatment. He focused carefully on her pattern of symptoms and course of disorder, although he showed relatively little interest in her relationship issues. He opined that they functioned as nonspecific stressors to trigger episodes but beyond that had no unique significance in explaining her condition. He diagnosed her as suffering from a bipolar

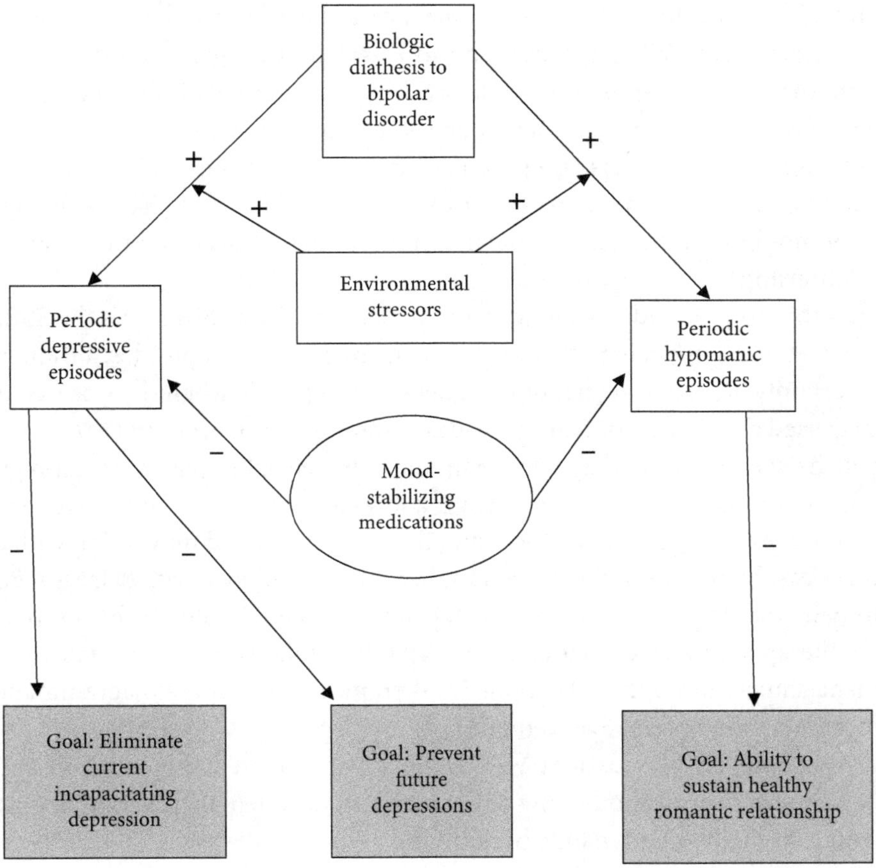

Figure 6.2 Mood disorder expert's inferred model for Ms M.

II disorder and treated her with a series of mood stabilizing medications including lithium, lamotrigine, carbamazepine, and valproate. During this time, she continued to be in and out of relationships with men and showed the same pattern of mood episodes. A diagrammatic representation of the expert's model might be as shown in Figure 6.2.

6.5 Another model failure and revision of the original model

After about a year Ms M felt there was no meaningful change in her symptoms and began to think that I might have had a better understanding of her condition. She then returned to me. She reviewed her experience with the mood disorder expert and the treatment trials involved as well as their lack of benefit. The continuing pattern of correlation between her mood episodes and her relationship status had become increasingly clear to the patient since I had originally called her attention to this. She was again in a depressed state owing to the ending of a relationship. She bemoaned how incredibly sensitive to rejection by a man she was. Upon further discussion she was able to describe how easily rejected she felt with even minimal disapproval from a man or anyone in an authority position in the workplace, although it did not lead to the catastrophic mood episodes that occurred in romantic relationships.

At this point I made two new recommendations. First, given that the SSRI had had little benefit in preventing subsequent depressive episodes, although it certainly helped bring her out of a depressive episode when she was in it, I suggested a trial of a monoamine oxidase inhibitor antidepressant (MAOI) as data exists suggesting they may be uniquely beneficial in patients with high rejection sensitivity[2]. Second, I again addressed the issue of intensive psychotherapy. I suggested that the issues that were generated in working with a male psychiatrist may simply be too disruptive to allow effective treatment to gain traction. I offered to help her find a highly qualified female psychotherapist with whom she could work while I limited my role to medication management and minimal supportive therapy. Figure 6.3 is a diagrammatic representation of this revised mode.

After extensive discussion, Ms M agreed with this recommendation and began intensive psychotherapy with an experienced female psychotherapist with a strong psychodynamic background.

Over the next two years the patient showed gradual and meaningful improvement in her condition. Rather quickly the MAOI appeared to give the patient some resiliency in the face of rejection. The psychotherapy made slow and steady progress. After about two years the patient established a healthy and enduring relationship with an appropriate man. They began living together and became engaged. There was minimal reemergence of her arrogant, haughty, and demanding attitude in this relationship, and when some of those features emerged the patient was able to recognize and control them with the help of her therapist and her fiancé, who at this point was sometimes included in the psychotherapy sessions with good benefit. She was able to sustain a consistent and realistically positive view of her fiancé that was neither overly idealizing or devaluing. Considering her marked progress, the patient raised with me the option of trying to discontinue her MAOI. I was struck by her ability to discuss this option as a collaboration between equals and not in the spirit of devaluing me. She stopped the MAOI without any reemergence of depression or worsening of her rejection sensitivity. She terminated treatment with me in an appropriate manner, but was planning to continue her psychotherapy work.

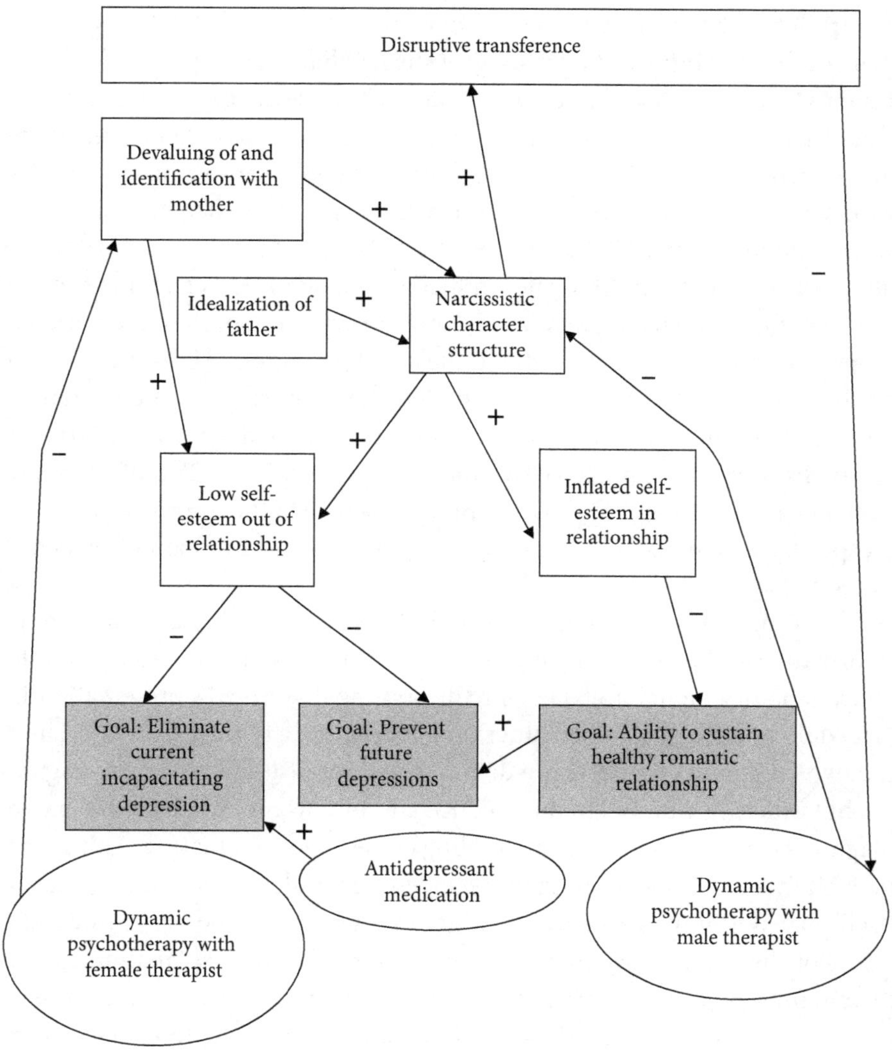

Figure 6.3 Revised model for Ms M.

6.6 Competing models: More than a disagreement about diagnosis

The case of Ms M provides further illustration of ways in which articulating instances of clinical reasoning in psychiatry in terms of a POP model adds clarity and value to the psychiatric treatment process. It would be easy to characterize the different viewpoints of the two psychiatrists involved in the treatment of Ms M simply in terms of a disagreement about diagnosis. Does she have a bipolar disorder or a unipolar disorder? Given the fuzzy edges of our diagnostic categories this could prove to be an insolvable argument. In such cases clinicians may be tempted to view response to drug treatment as the arbiter of diagnosis. Had she responded well to mood stabilizers, would that confirm the bipolar diagnosis, or if she had had an unequivocally strong response to antidepressants, the unipolar one? While drug response certainly would have some relevance, it is far from diagnostically definitive. Many mood stabilizers are known to have some antidepressant value in unipolar patients and bipolar II patients can in some cases show meaningful benefit with antidepressants. In the current DSM (*Diagnostic and Statistical Manual)*, symptom pattern dominates in making a diagnosis. In fact, I felt that if one were forced to make a diagnosis purely on symptom presentation and no other information about the patient, the case for a bipolar II disorder would be the stronger one. It was in looking at other features and circumstances involved with this case, factors that do not figure in the diagnostic criteria for either disorder, that an alternative model suggested itself. The POP model in essence is an attempt to provide some substance to the otherwise vacuous escape clause that occurs in so many diagnoses in DSM of 'cannot be better explained by another condition'.

A POP model is in essence a proposed explanation. A diagnosis is only informative to the extent that it provides elements that are incorporated into an explanatory model. I was essentially arguing that Ms M's presentation in fact could be better explained other than in terms of a bipolar disorder. Thus, the two competing models here demonstrate not just differences in diagnosis, but differing views as to the salience and importance of any DSM-based diagnosis to the proposed understanding of the patient's situation. Although I had felt that Ms M met criteria for recurrent major depression, the diagnosis itself played a limited role in the larger explanation of the patient's condition. For the second psychiatrist, diagnosis served as the explanation. Other life circumstances were reduced to nonspecific stressors that could serve to precipitate episodes of the disorder, but were not intrinsic to explaining it. Treating the disorder in his model is essentially synonymous with treating the diagnosis, while with my model, treating the depressive disorder had its place but addressing the larger pattern of Ms M's propensities was the critical issue.

At the risk of a bit of repetition, I want to highlight what is the most fundamental disagreement between the two psychiatrists here. In the current climate in psychiatry, too often we focus on the differential diagnosis as the central issue. However, that is not so here. The central issue is whether the diagnosis is the core explanation of Ms M's problems and whether that diagnosis suggests the most pivotal treatment options. The mood disorder expert felt it did. I suspect that if he had judged that the patient's symptoms fit better with a diagnosis of unipolar depression, the pharmacologic treatment of that condition with the appropriate medication would still have been seen as the pivotal treatment intervention and the vicissitudes of her romantic life still reduced to the status of non-specific stressors.

I know that as the initial psychiatrist, had I felt that a bipolar diagnosis was appropriate, my basic understanding of Ms M would have changed very little. I still would have felt that the dynamics of her narcissistic issues were the pivotal source of her distress and that intensive psychotherapy to address them was the key intervention. The only difference would have been selecting a mood stabilizer over a traditional antidepressant. Frankly, I suspect that the outcome would have been the same, as I viewed the pharmacological treatment of any underlying mood disorder as of secondary importance in this case. My decision not to give her a bipolar diagnosis mainly reflected my belief that the psychodynamics adequately accounted for her pattern of symptoms. The hypothesized character structure could be explained in terms of the very powerful and specific dynamics in the relationship between the patient, her father, and her mother. It not only explained the intense mood shifts accompanying the dramatic fluctuations in her sense of self-worth, but also explained their coinciding with the state of her current relationships with a man. There was nothing left to explain by embracing a bipolar diagnosis.

6.7 A major decision: To revise or jettison a failed model

Yet there is something else that is left to explain. If my psychodynamic understanding was essentially correct, why did my efforts at conducting psychotherapy with Ms. M based on it fail, and why did the therapy of the subsequent therapist, who had the same basic understanding of the dynamics, succeed? When a prediction about the patient's response to a treatment intervention fails to occur, it speaks to some inadequacy or inaccuracy in the POP model. After all models are to be judged by their success in generating accurate predictions. I originally predicted that psychodynamic psychotherapy conducted by me would allow the patient to come to an understanding of the counterproductive patterns in her relationships with men and to use that understanding to alter both her sense of self and her capacity to sustain a healthy relationship. Instead, she disrupted the therapeutic work multiple times and eventually stopped therapy with me altogether.

When she returned to treatment with me after her time with the other psychiatrist, I was forced to rethink the model given this failed prediction. In this sort of situation, a clinician faces a major decision point. One option is to say that the model is fundamentally wrong, to jettison it and start again from scratch to generate an entirely new model. The second option is to hypothesize that the original model is in its essential features correct and can be salvaged with more modest adjustments of the model. There are no unambiguous guidelines here. And there are risks with either choice. In taking the first option, I might be throwing out an essentially correct model that could be salvaged with relatively modest modifications. With the second option, the risk is that I might explain away almost any failed prediction by adding more and more ad hoc modifications. This dilemma has interesting parallels with the well-recognized problem in the philosophy of science of whether to abandon the theory based on some anomalous outcome of an experiment or to propose an alternative explanation that leaves the major elements of the theory intact.[3]

In this case I opted for the second approach. This was based on several considerations. In my work with Ms M, she was able to see the issues quite clearly and to accept their implications when she was depressed and not in a relationship. She appeared to show real insight and made enough progress that her mood began to improve even without the beginnings of a new relationship with a man. But when she began a new relationship, she rather quickly became haughty and arrogant, and increasingly more dismissive of me and my interpretations. Hence the way my treatment of her failed was precisely through a reenactment of the dynamic pattern that was at the heart of my model. This led me to hypothesize that the essentials of the model were correct, and that the error lay in overestimating my capacity to successfully address the pattern of idealization alternating with contempt that emerged in all her relationships with men, including me. I overestimated her capacity to maintain the insights she gained when depressed, that is to sustain an observing ego, as she became more grandiose. This modification of the model was essentially a quantitative one, in that I thought that the strength of the therapeutic forces in psychotherapy with me would be able to overcome the idealization/contempt cycle. I came to believe that was not the case.

Based upon this modified understanding, I hypothesized that psychotherapy based upon the same basic set of issues could be successful if conducted by a woman, thereby sidestepping her counterproductive response to men. This proved to be the case, lending support to this modified model.

6.8 Using the transference to generate predictions

While the basic framework for understanding Ms M remained intact in the revised model, many more complexities emerged from our relatively brief psychotherapeutic work together from the time when she returned to treatment with me to when she transferred her care to a female therapist. While at the most obvious level my psychotherapeutic work with her, based on my model, had failed, I was struck by the fact that she returned to me at all. How could I understand this and what light did it shed on how to revise the model? Ms M described her decision to leave the other psychiatrist both because she felt his treatment approach was not helpful, but also because she saw him as arrogant and dismissive of her. I pointed out to her that her failure to make the psychiatrist value her resembled her experience with her narcissistic father whose respect she could never win. She was able to see the parallel, and commented that even during times when she saw me as ineffectual, she always felt I tried to understand her and listened to what she had to say.

In this light we again reviewed her prior romantic relationships with men. She talked about the fact that she tended to choose men that she felt she might not be able to win, but that once she won them and she became haughty and arrogant towards them, she would take any effort on their part to work with her and temporarily accept her bad treatment of them as a sign of their weakness and inferiority. As a result, she would treat them in an even more demeaning manner, until they were eventually driven to leave her. When she then became depressed, she could look at their attempts at acceptance in a more positive light and regret the loss of the relationship.

I pointed out to her that she was returning to me at the time when she was depressed, and that I represented one of those accepting but devalued men. But unlike her romantic involvements, the therapeutic relationship was one to which she could in fact return. The question arose as to whether we should try psychotherapy again. What appeared like further growth and insight made that a tempting option. However, based on our past experience, I had deep doubts as to whether she could sustain workable insight with me when her depression lifted. In my own mind I spent some time struggling with the possibility that, in not continuing to work with her, I was responding to my own countertransference of wishing to get rid of her before I was devalued and left once again. However, I concluded that the chances of success with me in psychotherapy were slim. I was very concerned however to make sure she did not experience my suggestion of transferring to a woman as a rejection of her.

These considerations led to a final refinement of the model. I realized that I had been focusing primarily on the need to address her pattern of idealizing/devaluing men. Perhaps her basic devaluation of herself as a woman, reflecting her father's general devaluation of women as well as some identification with her devalued mother, was the more central issue. Perhaps she would be unable to feel good about herself without a man, and a sufficiently arrogant one at that, unless she could first accept that a woman could be intrinsically valuable and worthwhile. I shared this perspective with her, and after some discussion, she felt that this interpretation was accurate. Based on this understanding I suggested to her that I felt her best chance of success in therapy, particularly once her depression lifted, was with a strong and competent woman, in contrast to her weak and ineffectual mother, who could help her appreciate her own intrinsic worth unrelated to any connection with a man. This approach made sense to her and ultimately proved quite successful.

6.9 Clinical patience is required to run a natural experiment

It took a long time for full clarity to emerge in this complicated case. Nonetheless, this case demonstrates that there should be, with any POP model, some relevant empirical data that will eventually allow some assessment of the adequacy of the model. In short, the model should not just provide an explanatory framework for the present situation of the patient, it should generate some predictions. In clinical work one must often wait for nature to run the experiment, but ultimately, for the model to be meaningful, differential outcomes that argue for or against the model are needed. This is an important component of any empirically grounded discipline.

A long time passed before the evolution of the patient's condition could reflect on the adequacy of my original model as well as demonstrating the need for some modification of it. However, as noted earlier, other features had emerged previously that added credibility to that model. Once the clinician has developed a POP model, further information gathered about the patient's past over the course of the treatment relationship is also at times a form of prediction. New information about the patient's past can be assessed in terms of whether it would be expected by the psychiatrist's POP model and hence either confirmatory or conflicting with what the model would expect and hence challenge the adequacy of the model. Obviously much added history will be neutral with respect to the model and not particularly informative. Nevertheless, the natural experiment which is the patient's life has been going on for considerable time before the psychiatrist first meets the patient, and it is to be expected that if the model is truly showing patterns of propensities that influence the patient's life, evidence that they are operating should be common.

Endnotes

1. My proposed model has much in common with Alice Miller's work on the bipolarity of the narcissistic character (Miller and Ward 1981).
2. Rejection sensitivity is viewed as the core feature of a pattern called atypical depression and appears in some studies to show a preferential response to MAOIs. Other features often associated with atypical depression (although how well they hold together as a cluster is controversial) are hyperphagia (excessive eating), hypersomnia (excessive sleeping), and profound lack of energy and motivation termed 'leaden paralysis'. Ms M has all these features when depressed (see Davidson 2007; Singh and Williams 2006; Gaynes 2009; and Pae *et al.* 2009). A similar sort of pattern was described earlier in the literature under the rather sexist term of 'hysteroid dysphoria' (see Liebowitz and Klein 1979, and Spitzer and Williams 1982)
3. Kuhn 1970; Lakatos 1978

Chapter 7
The POP model of clinical reasoning

Description and implications

7.1 Three roles for POP models in psychiatry

We have looked at a series of specific case studies that concretely illustrates how a pattern of propensities (POP) model works as a clinical method of reasoning. Our task now is to step back from these specific cases to formulate a more general picture of what POP modelling is and its implications for understanding the practice of psychiatry. In this chapter, then, we will take a more comprehensive view of the hypothesis that is the main theme of this book; namely, that we can understand the methodology of clinical reasoning in psychiatry as practised by competent psychiatrists as a process of model construction specific to an individual patient. Such a model consists of a collection of propensities and the interactional links between them organized around the goals for which the patient seeks help. A POP model offers a hypothesis to explain how the patient's current situation developed and can generate a set of predictions as to the impact of potential interventions. While effective models may differ based on the background and training of the individual psychiatrist, for any initial model the strength of the empirical support for each propensity and the interactions between them should be carefully considered before interventions derived from the model are implemented. The generalized knowledge base of psychiatry is drawn upon for this. This is a conceptually different and more fine-grained use of evidence than the approach advocated by evidence-based medicine (EBM). The model, as hypothesis, is then revised based upon the further evidence gathered as the case unfolds over time.

POP modelling is a theory about the methodology of clinical reasoning in psychiatry that is meant to operate on three levels:.

1. As a **description**. POP modelling is not a new way of clinical reasoning. Rather it is an explicit articulation of what good clinicians have long done tacitly. Elaborating and characterizing this method of reasoning

How Psychiatrists Make Decisions. Douglas W. Heinrichs, Oxford University Press. © Oxford University Press (2025).
DOI: 10.1093/9780198960720.003.0007

will allow identification of several key features of POP models. In so doing, it will promote self-understanding for the clinician, and draw out several important implications about the nature of psychiatric practice, what activities are at its core, and how it relates to several closely related, but quite distinct, other disciplines. Fleshing this out will be the primary focus of this chapter. Some of these features of POP models have been discussed in our consideration of the individual cases, but we will generalize and summarize them here.

2. As a **pedagogical tool.** By making the clinical reasoning process explicit, POP modelling encourages critical reflection on the process of clinical thinking, facilitating a more accurate assessment of the quality of that reasoning. This allows clinicians to better evaluate their own work, reduce the impact of problematic potential biases, and self-correct and improve their clinical thinking. It promotes the thoughtful and targeted application of more general knowledge to the individual patient, such as that derived from clinical trials, naturalistic studies, basic science research, and collective clinical wisdom. It also facilitates effective collaboration between multiple clinicians involved with a patient, which is especially crucial when there is some difference in viewpoint about the optimal treatment approaches. These same features of the POP modelling approach can be useful in optimizing the training of new clinicians. In essence, it is a vehicle that encourages the psychiatrist's progression from the status of a novice to that of an expert. POP model construction is a rationally justifiable process totally in keeping with what recent cognitive research has shown to be the process of developing true expertise in any field. Considering these very pragmatic, pedagogical implications of the theory will be the focus of Chapters 8 and 9.
3. As an **epistemological justification**. I argued in section 1.3 that by viewing clinical reasoning about an individual patient as simply an intuitive or impressionistic activity, i.e. the 'art' of medicine, the value of that activity has been steadily eroded in an era where formal research trials with their statistically supported conclusions are viewed as the benchmark of good science, and science is seen as the best standard for credible knowledge. By articulating the process of clinical reasoning, we may critically assess the epistemic status of this way of thinking and its relationship to scientific method. The understanding of scientific method that has evolved in recent decades in the philosophy of science supports the position that model construction, of which POP

modelling in psychiatry is an instance, is in fact the paradigm of scientific method as it is practised in all fields of science. This will be the focus of Chapter 10.

7.2 Can we study how experts think?

A central rationale for this book is that clinical reasoning has traditionally been an implicit and seemingly intuitive process, with the actual method typically outside of the awareness of the clinician. How then can a theory of that method be anything other than wild speculation as to what is going on in the mind of the clinician? Fortunately, there is a substantial and growing body of research by cognitive scientists, prominent among them Gary Klein[1] and Gerd Gigerenzer,[2] on the development of expertise in a wide range of fields. Klein has developed structured interviewing techniques that help experts discover for themselves an explicit understanding of their previously tacit judgements. They have looked at firefighters, military squad leaders, and emergency medical personnel, among others.

An example from Klein's work will illustrate this process.[3] A lieutenant leads a group of firefights into a one-story home. The fire is in the back, thought to be contained in the kitchen area. Several attempts to hose water on the blaze have surprisingly little impact. The lieutenant, for no conscious reason, senses that something is not right, and orders his crew to leave the building. Immediately after exiting, the floor where they had been standing collapses. Had they been in the building, they would have plunged into the fire below. Initially the lieutenant could only appeal to his 'sixth sense' acquired from years of experience. On structured questioning to carefully reconstruct the sequence of events as he experienced them, the lieutenant was able to explicitly identify several cues and his processing of them. While he initially had no reason to suspect this was a basement fire, he was surprised that attempts to hose water into the kitchen had so little effect. Then he remembered thinking that the house was unusually hot for a small kitchen fire. Finally, the fire was unusually quiet. His intuition now made rational sense to him. Hosing water into the kitchen had little impact as the fire source was in the basement. The burning basement heated the entire house in a way a kitchen fire would not. Finally, the floor muffled the noise of the fire below.

As experts make critical decisions, especially in conditions of extreme time pressure and high ambiguity, they often experience their judgements as intuitions for which they can at first provide very little articulation as to their reasoning. Using specialized research interviewing techniques, experts can come to identify the key elements of their reasoning process that are behind their intuitions. Individuals can learn to apply many of these strategies on their own in introspecting about their own reasoning processes. As will be examined extensively in Chapter 8, the pattern of reasoning discovered by Klein employing this procedure bears striking parallels to that used in POP model construction. For now, the important point is that it is possible for experts in a field to discover consciously the method by which they arrive seemingly intuitive judgements.

7.3 Using models in the context of other psychiatric skills

Before attempting to characterize the general features of POP models, it is important to re-emphasize how clinical reasoning and decision-making fits into the larger framework of what psychiatrists do and what skills they draw upon. As discussed in Chapter 1, psychiatrists' training and accumulated experience lead to two sorts of data-gathering skills presupposed in constructing successful models. First, they learn how to acquire and update a body of general knowledge, including about groups of patients with certain features in common, such as diagnosis. This knowledge can come from multiple sources. Much of it comes from the published literature including both reports of formal research studies as well as more descriptive writings reflecting generalizations from accumulated clinical experience. Other research involves basic science mechanisms with relevance to clinical issues. Other sources of knowledge include advice and sharing of experience from supervisors and teachers while in training and from colleagues subsequently. Over time psychiatrists themselves acquire a growing body of experience from which certain generalizations can be made.

The second skill required by psychiatrists is a range of techniques for systematically collecting and organizing a vast amount of unique data from an individual patient. These include obtaining a comprehensive history, conducting a careful mental status examination, and developing a safe and trusting therapeutic alliance with the patient that facilitates ongoing sharing and communication.

The component of the psychiatrist's work that constitutes clinical reasoning and decision-making, and that I argue is best captured in terms of generating a POP model of the patient's situation, involves how to take this vast body of both general and specific information relevant to a particular patient, and deliberate about it in a way that yields a framework—a model—that both suggests a meaningful explanation of how the patient's situation came about and, most importantly, suggests a rational basis for making predictions about how the patient will respond to future experiences, especially possible treatment interventions.

7.4 The domain of a POP model is a unique individual

What then are the important general features of POP models in psychiatry? To avoid confusion, it is important to distinguish between the model as an abstract, cognitive construct and the way that model is represented and communicated. The latter can take many forms for the same model. Throughout this discussion, models have been most often expressed in narrative form—described with a series of sentences constituting a kind of story that articulates propensities and their interactions. It is equally possible to use other formats. To illustrate this, at multiple points in our discussion I used box and arrow diagrams to capture the essential structure of a particular model. I personally find that narrative representations lend themselves to more nuanced discussions of the variables involved, whereas diagrams can make it easier to see the overall pattern and structure of the model. But the content of the model is independent of the way it is represented.

An important starting point to understanding POP models is to recognize that they are always about a specific concrete patient, not a group or category of patients. Every scientific discipline has its paradigmatic context in which it operates. Although psychiatry is applied in many types of settings, the paradigmatic context is the psychiatrist engaging with an individual seeking help for a specific form of suffering or dysfunction. POP model construction is intended to operate specifically at this level of the individual.

This is its most foundational difference from approaches like EBM, where the paradigmatic context is data generated from studies of groups of patients subjected to statistical analysis to yield probabilistic predictions that then are applied to members of that group. As a result, EBM is a large step removed from the sorts of questions asked by the psychiatrist about an individual patient. In response to this disconnect, EBM offers several methods that are intended to allow clinicians to work from the group data back to their individual patients. As we have seen in Chapter 2, this is not very successful.

An alternative approach to this problem is model construction. Here the clinician attempts to articulate an understanding of the processes at work in individual patients that have led to their difficulties, leading to the identification of points of potential intervention to alleviate their suffering.

The psychiatrist does this by postulating a number of elements, termed propensities, and the pathways of influence linking these propensities, thereby providing a theory of what led to the development of the patient's difficulties and what points for intervention might be available. Information from group data is certainly used, but rather than being applied statistically, it is incorporated into a hypothesized pattern of interacting propensities intended to capture the uniqueness of this patient's situation. Data from group studies are not used to answer a question in the form of 'What is the best treatment for a patient like Ms Jones'? Rather it addresses a question in the form of 'To what degree does this data support the likelihood of a particular propensity having the proposed influence on Ms Jones' condition suggested by our model'? In a similar vein, a diagnosis is useful to the extent to which it suggests certain propensities and their interactions that may be relevant components of our model of Ms Jones.

Like a map, a model is a description of a concrete piece of reality, a portion of that universe that is constituted by this unique individual patient. If such model construction is a scientific activity, an issue that will be taken up more extensively in Chapter 10, then that activity is not the study of schizophrenia, or bipolar disorder, or depression or anxiety. Rather, when we are treating Ms Jones, we are practising something best seen as 'Jonesology' and when we are treating Mr Smith, 'Smithology'.

7.5 Distinguishing psychiatry proper from its informing disciplines

But when a research psychiatrist is conducting a clinical trial, isn't that activity psychiatry? Or when a practising psychiatrist is reading a study in a psychiatric journal, aren't they learning psychiatry? No doubt the knee-jerk reaction of anyone in the field is to answer yes to both questions. But perhaps some reconceptualization is in order. I suggest what may seem like a rather radical extrapolation from the model theory of clinical reasoning, but one that reduces some of the conceptual confusions about the appropriate methods for the practising psychiatrist. If the paradigmatic activity of the psychiatrist is to generate a model of the individual patient in order to develop a specific understanding of that patient's circumstances and potential avenues for intervention, then when engaged in that activity the psychiatrist is practising psychiatry proper. In the process of generating a model, however, the psychiatrist draws on a range of information from other disciplines that are highly informing for the practice of psychiatry. In some cases, it is easy to see these disciplines as something distinct from psychiatry itself. Insights from basic biological sciences such as neuroscience, biochemistry, pharmacology, molecular genetics, and cognitive science are highly relevant to psychiatry but distinct from it, with distinct methods appropriate to each discipline. This is also true for psychological and social sciences that may provide useful insights for the practising psychiatrist.

I would suggest that there are at least two other disciplines that, while more intimately connected with psychiatry, are nonetheless best conceptualized as distinct from it: psychopathology and clinical therapeutics. The former deals with the characterization of psychiatric signs and symptoms as well as the subjective experiences of groups of similar patients, often but not necessarily defined by diagnosis. The latter deals with assessing treatment interventions in groups of patients sharing certain common features, again usually a diagnosis. These two disciplines may be more difficult to view as something different from psychiatry proper. In large part this is because they are usually conducted by individuals trained in psychiatry. I am not arguing that this is inappropriate, but not everything a psychiatrist does needs be psychiatry proper.

One of the difficulties that comes from confounding these disciplines with psychiatry proper is the assumption that the methodologies used in these disciplines are essentially the methodologies that should be used by the practising psychiatrist as much as possible. In fact, the psychiatrist may be viewed as only being scientific when using these methodologies,[4] with anything else at best seen as a stopgap for the limits of our scientific knowledge, the 'art' of medicine. The proposed relationship between psychiatry and its informing disciplines is illustrated in Figure 7.1.

7.6 POP models are built around the patient's goals

Let us return to a consideration of some of the general attributes of POP models. All models are constructed with a specific purpose in mind. Thus, the first step in model construction in psychiatry is to identify the goals of treatment. For what is the patient seeking help in consulting the psychiatrist? Is it sufficiently well-characterized to provide a useful starting point for model construction? Is it realistic? Can a goal be constructed that can be shared by both the patient and the psychiatrist? In many cases this is fairly straightforward, while in others considerable reformulation and negotiation with the patient is required. Thus, for Ms K and the initial treatment goals for Mr Z, a shared understanding with relatively little negotiation was easily

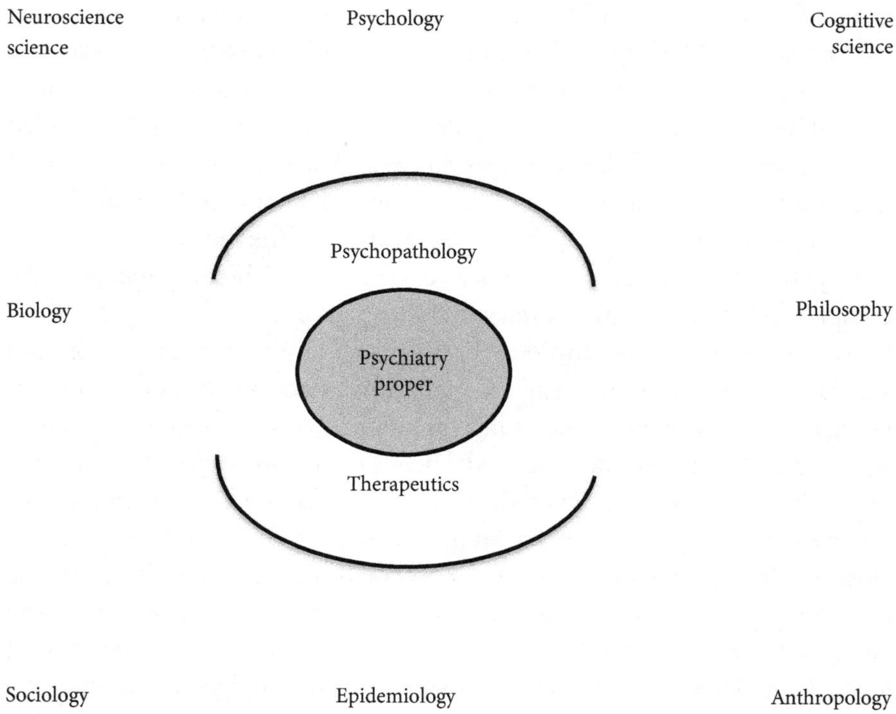

Figure 7.1 Psychiatry proper and its informing disciplines.

Reproduced with permission from Heinrichs, D. 2015. Model-based science and the ethics of ongoing treatment negotiations. In J. Z. Sadler, C. W. Van Staten, and K. W. M. Fulford (eds). *The Oxford Handbook of Psychiatric Ethics*. Oxford, UK: Oxford University Press.

reached between doctor and patient. However, for the later modified goals of Mr Z and certainly in the case of Mr D extensive changes by both doctor and patient were needed to reach a common ground. This requires a negotiation process between doctor and patient that is in fact highly useful. This process itself is a large step toward establishing a working collaborative alliance with the patient. It not only requires the patient to look at their goals carefully, but requires psychiatrists to consider whether they really understand their patients' goals or are imposing some preconceived or formulaic goal such as 'this patient suffers from depression and therefore my goal is to reduce depressive symptoms'. This by itself is unlikely to capture any nuances of the patient's particular circumstances and form of suffering. Furthermore, as we have seen in several of the cases, the patient's specific goals may significantly shift over time and with changes in circumstance, and at times can be highly idiosyncratic.

It follows from this central role for goals that the meaningful test for the adequacy of a model is its ability to lead to interventions that help to meet the goals. No model is a complete description of the piece of reality it is intending to characterize. Many possible factors are left out because they are judged to be of little or no relevance to the task at hand. Again, the analogy of a map is helpful here, as we discussed more fully in section 3.8. The map is intended to characterize a particular geographical area. That area can be characterized in many different dimensions and no one map can contain them all. Which ones need to be included depend on the purpose of the map.

The challenge for the clinician is to determine which components of the complicated reality are in fact relevant for the patient's goals. Here one must draw on the generalized knowledge base gained in training from mentors and supervisors and from following the literature, as well as one's own accumulating clinical experience. The same argument applies to the question of how accurate and well calibrated the model needs to be. This cannot be answered in the abstract. Of course, no model is perfectly accurate as all measurements involve some margin of error. For some purposes a map needs to be drawn as closely to scale as possible, whereas for others the map can be quite imprecise and still get the job done. Maps and models are very pragmatic things. The adequacy of the POP model, both with respect to what is included at what level of detail and how precise it needs to be, is to be judged by whether the model serves its purpose of generating a rational and effective treatment plan for the patient.

7.7 The elements of a POP model are propensities

The type of model used in clinical psychiatry is composed of elements termed propensities and pathways of influence between them. In many disciplines, especially the hard sciences such as physics, patterns are typically seen as made up of elements that function as causes with the interaction between them seen as that between cause and effect. I will argue that propensities are in fact the best way to conceptualize causes in all scientific disciplines, even in physics. The rationale for this position is best considered as part of a more extensive consideration of the nature of scientific method that is taken up in section 10.2.3.9.

Here I want to emphasize two reasons to prefer the term propensities and pathways of influence that are especially relevant to psychiatry. The first is that for many people cause and effect imply totally deterministic models that leave no room for notions of human freedom or people doing things based on reasons as opposed to causes. While issues of human freedom and free will are contested philosophical notions, there is no reason for psychiatric models to take a stand on these questions. Hence it makes sense to choose a terminology that leaves open the ontological status of the elements of the model and in what sense they influence one another. Furthermore, advocates of psychiatric models need not take a stance on the question of whether doing something for a reason is categorically distinct from causation or whether reasons are simply a particular type of cause. Similarly, the term mechanism will be generally eschewed in favour of patterns of influence, to avoid implying any stance on the issue of rigid determinism.

7.8 Propensities are conditional assertions

An additional reason to choose the notion of propensities over causes is to emphasize that propensities imply conditional assertions. Thus, with Ms K, the model is not asserting that she is always anxious. Instead, she has a propensity such that faced with a novel situation or the threatened loss of people on whom she is dependent for her sense of security, then she is likely to become highly anxious. Or consider the assertion for Mr Z that if he is made to feel humiliated or emotionally abused, then he will become more detached and withdrawn, and if the humiliation and abuse continues, he is likely to lash out with extreme anger. Even more clearly in the case of Ms M, the assertion is that if she is in a relationship with a man then she will become increasingly grandiose and demanding, and if that relationship ends, she will become clinically depressed.

This conditional quality applies to all cause-and-effect relationships. So, only if you drop the cannonballs from the leaning Tower of Pisa will they fall to the earth with a particular acceleration. If you do not drop them, they will not fall. If you mix hydrogen and oxygen, you will get water, whereas if you do not mix them, you will not. However, in a high percentage of cases in the 'hard' sciences, this conditional aspect is so self-evident as to hardly need mentioning. In psychiatry however these conditionals are pervasive and not always self-evident. Hence, the relationships between propensities may not appear to operate in a precise or law-like fashion. Rather, their expression is context dependent.

Many of the features that we think of as traits or characteristics of an individual are not features that are always in evidence, but rather represent dispositions or propensities to behave or feel in certain ways when faced with certain circumstances. Since the specifics of such circumstances are typically so complex and variable, this is part of the reason that when we try to formulate predictions, especially about groups of individuals, the results are so probabilistic. However, it is important to understand that the probability of an event occurring is not a measure of the strength of that propensity in any consistent way. Thus, an acorn has a strong propensity to develop into an oak tree. However, most acorns in fact do not develop into oak trees. The use of the term propensities in model construction serves to keep this issue salient.

7.9 An inverse relationship between richness of context and precision of predictions

In a field like psychiatry patients are always embedded in complex and ever-changing contexts. This contributes to making precise predictions impossible as well as precluding precise quantification of propensities and their interactions. Part of the power of fields like physics derives from the ability to isolate elements in artificial circumstances so that precise interactions between a few elements (variables) can be carefully measured under highly controlled conditions (that is, minimizing the effect of confounding variables). As a result, law-like relationships can be established, and where measurement is sufficiently precise those relationships can be expressed as equations that can be manipulated successfully for many purposes. These kinds of contrived and simplified circumstances have been termed 'nomological machines', from the Greek word *nomos*, or law, by the philosopher Nancy Cartwright.[5] It is only a contrived condition with a small number of highly controlled elements, shielding off all other external influences, that can generate something that can look like a scientific law.

Ironically, when such a law-like relationship is established by use of a nomological machine, it never literally applies to any real-world situation. Sometimes real-world conditions can be made sufficiently constrained that these law-like relationships hold close enough to describe those conditions adequately. Even more infrequently certain natural systems may be sufficiently simple and isolated to allow these law-like relationships to describe them with some accuracy. The application of laws of gravity and laws of motion to the behaviour of planets in the solar system is a striking example.

However, even in physics, if the elements are embedded in a sufficiently rich and diverse context, precise calculations are impossible. Cartwright, citing the philosopher Otto Neurath, uses the example of releasing a crumpled dollar bill on a windy day in St. Stephen's Square and then trying to calculate where it will end up.[6] She argues in fact that the issue is not that the equations would be too complex to calculate, but that in this setting they simply do not exist. Similarly, the nature of the subject matter of psychiatry is such that nomological machines simply do not exist; the elements and variables involved are countless. Hence precise quantification of relationships between elements is impossible in theory.

7.10 In what sense are POP models quantitative?

Having said this, psychiatric models are not totally devoid of a quantitative dimension. It is simply a very crude one. When positing an influence path between two elements in a psychiatric model, the hypothesis includes a prediction as to whether element A will increase or decrease element B. Often there will be multiple elements impacting on element B, with some increasing and others decreasing it. The clinician must speculate on the likely net effect of these factors without being able to precisely measure them or to calculate their impact with an equation. While clearly less precise than a model from physics, a POP model in psychiatry nonetheless generates a prediction. Whether that prediction holds goes a long way to determining the adequacy of the model and whether the model should be maintained, modified, or rejected. An important model revision in the case of Ms M was a response to a quantitative failure of my prediction. I had vastly underestimated the power of her transference pattern of devaluing me when in a relationship and her lack of insight when that occurred. I also overestimated the power of our therapeutic alliance for working through this issue without a disruption of treatment.

An additional reason, of course, why psychiatric models will be less precise than models in physics is that the elements involved cannot easily be assigned meaningful numerical values. When we measure things like temperature, mass, and velocity there is a literal meaning to the assigned numbers and the relationship between these quantities can be precisely and meaningfully characterized and manipulated. This is not so with the kinds of elements that make up a psychiatric model. (An exception is when a model incorporates pharmacokinetic or pharmacodynamic variables such as strength of receptor binding or competitive inhibition by another drug.)

We frequently see attempts to circumvent this problem with the use of rating scales that in fact do assign numbers. As discussed in section 2.6, these numbers must be looked at very critically as they are often instances of pseudo-quantification. If a depressed patient moves from a six to a three on a depression rating scale, is he literally only half as depressed? Has that patient become less depressed in the same magnitude as a patient who is moved from a nine to a six? Or if one patient moves from a nine to a three, has the amount of depression in the world been reduced by the same amount as if two patients each move from a nine to a six? These questions sound silly precisely because the assigned numbers are not truly measurements at all. Hence, I would argue that in psychiatric model construction it is more honest to keep quantitative predictions at the level of increasing or decreasing, perhaps at times adding 'by a lot' or 'by a little'. This is likely as precise as we can honestly get. Nonetheless such quantitatively crude predictions are often adequate to generate a hypothesis about treatment and to test its efficacy for a given patient.

What is clear from this use of the concept of propensities is that, while we may postulate persistent traits and attributes as elements in a POP model, what is most often being explained and predicted are patterns of experience and behaviour as opposed to fixed states. That is, given some set of circumstances and conditions, certain types of actions or subjective experiences are likely to occur in this patient. In some ways this adds to the predictive strength of the model, in that each time that set of circumstances occurs a new prediction follows. In each instance its occurrence or non-occurrence adds to or subtracts from the confidence we can have in the model. Thus, in the case of Ms M, the fact that her pattern of response to entering and then losing relationships had repeated so many times and so consistently led me to have heightened confidence in the model I had proposed.

7.11 Models integrate multiple levels of explanation

Another noteworthy feature of POP models is that the elements in any given model can be drawn from a full range of levels of explanation. A huge amount of ink has been spilled over the years debating the question as to the most appropriate level at which to explain mental and psychiatric phenomena. At the one end are biological reductionists who believe that with sufficient scientific advancement all mental phenomena can be best explained in terms of simple biological variables—the 'biological' level. Even among these individuals there is considerable debate about how granular that level of biological explanation should be: at the molecular level, the genetic level, the level of neural circuitry, neurochemistry, or the like. At the other end are those who would argue that mental phenomena are fundamentally irreducible to biology, and must be explained in categories unique to the psychological sciences. Among these individuals are some dualists who believe that mental phenomena are ontologically different from physical phenomena and hence require different levels of explanation. Others, now in the majority, emphasize the role of emergence, that is the notion that while all mental phenomena have physical correlates, as physical elements and their interactions become increasingly complex, unique properties emerge that cannot be explained in terms of the more basic levels of physics and chemistry.[7]

Intimately tied to this discussion of different levels of explanation are notions of different processes by which we understand nature. An important distinction dating to the late 1800s in certain schools of German social science, and brought to prominence in discussions of psychiatric explanation by Karl Jaspers,[8] is the distinction between *erklären* and *verstehen*. The former is usually translated as 'explaining', and refers to postulating a connection based on traditional cause and effect, supported by empirical observations and studies using what is traditionally viewed as scientific method. The latter, usually translated as 'understanding', refers to the human capacity to directly perceive the meaningful connection between elements. Thus, an individual who is treated in an unfair or humiliating way is likely to respond with anger. We, as fellow humans, directly experience the meaningful connection between the mistreatment and the anger so that the angry response is directly understandable to us without the need for accumulating extensive empirical data showing that mistreatment is followed by anger. Mr Z's pattern of response to his mother is an example. We find it understandable without needing further empirical data.

Biological explanations by necessity represent knowledge derived from *erklären*. Psychological explanations could derive from either *erklären*, as in experimental psychology paradigms, or from *verstehen*, or both. Thus the tendency for a man to find a female more sexually attractive if he encounters her in an anxiety-provoking situation, was not generally seen as intuitively understandable before it was demonstrated in an experimental situation with quantified data and hence is an example of *erklären*.[9] That a man finds a particular woman attractive because she looks very much like his first love is something we can understand as a fellow human being without asking for empirical studies to support it. The fact that explaining and understanding are often treated as interchangeable in ordinary discourse makes it more difficult to keep this distinction in mind. Nevertheless, I find this to be a useful distinction in considering the full range of processes by which we can acquire knowledge. It is problematic however, as elaborated in section 10.2.1.2, if the distinction is used as a basis for separating what is scientific from non-scientific.

For all the controversy these issues generate at the theoretical level, in the practical task of model construction psychiatrists can freely draw on elements and connections from any level of explanation that is pragmatically useful in the given case. Moreover, they can freely combine those elements together in any given model without contradiction. The postulated interactions between elements can be understood either within the framework of *erklären* or *verstehen*. The real issue is which level of analysis is most practically helpful for understanding and manipulating an element that is deemed important in generating the patient's difficulties. In fact, it is obvious in many cases that the same element or path of influence between elements could be characterized on several different levels depending upon the practical need at issue.

We are dealing with a methodological, not an ontological, contrast. Thus, in the case of Mr Z the first formulation of the model argued for the value of an antidepressant. This was based solely on empirical studies showing a benefit for antidepressants in many cases of patients who had symptoms of major depressive illness, which Mr Z did. It was unnecessary at this point to cast the model in any more biologically fine-grained terms such as the specific neurotransmitters involved in the antidepressant's action. However, as the patient's issues became more convoluted, speculation as to possibly differential effects of various types of antidepressants became important in the model. This required me to employ levels of explanation that included specific neurotransmitter action that would be hypothesized to predict potential differential effects among antidepressants. There are no *a priori* privileged or preferred levels apart from the practical needs of the model and the background knowledge of the clinician.

7.12 Prediction promotes rigour: Pluralism versus eclecticism

The tremendous flexibility available to psychiatrists in constructing a model may well raise some red flags. Is this just a fluid, anything-goes process where clinicians can throw any speculation into their explanations of the patient's circumstances and proposed treatment? Is this simply eclecticism run wild? Nassir Ghaemi[10] has provided a detailed and thoughtful discussion of the problems of eclecticism in psychiatry. He argues that in response to a prior era in psychiatry characterized by extreme dogmatism on the part of both psychoanalytic and biological thinkers, psychiatry attempted to bridge this gap by embracing all approaches under the rubric of the biopsychosocial model originally proposed by George Engel.[11] As Ghaemi rightly observes, the biopsychosocial model in fact is not a model at all. It is simply an admonition that factors on biological, psychological, and social levels should all be considered for each patient. While hard to argue against, this instruction is vacuous and short on particulars. It gives no guidance in how to decide in any given case which elements at which levels are most relevant or important.

Instead of eclecticism, Ghaemi argues for what he terms 'pluralism'. He sees pluralism as the willingness to utilize understanding and treatments drawn from various levels of analysis, but based on discrete predictions about which specific treatments are likely to be most helpful in specific circumstances. By this he means not just choosing between psychotherapy and pharmacotherapy. He also means selecting a specific psychotherapeutic approach in each patient, or specific type of pharmacological intervention. It also requires deciding when it is important to use certain families of modalities and when it is not. I think that Ghaemi's analysis of these issues is quite accurate and very much in keeping with the project of POP modelling.

I would argue that POP model construction represents an instance of pluralism at work as opposed to eclecticism. In introducing a specific element or positing a path of influence between elements in constructing a model for a specific patient, the psychiatrist must commit to a very specific kind of prediction that leads to a particular proposed level of intervention. By hypothesizing how one element affects the next, any proposed treatment modality that influences that pathway is being posited to have a specific effect on the process. Thus, in the case of Mr Z, at various points in his treatment different forms of psychotherapy are provided. Early on, exploratory psychotherapy was used and proved useful because it helped the patient identify and address patterns from his early life that so deeply impacted his current relationships. Later more supportive psychotherapy was needed to help him weather the turmoil of the ending of his marriage. Here the goal was helping him contain his affect and modulate his responses. How was this later choice of psychotherapy modality made? I drew on both my own experience and prior guidance in my training and clinical supervision to postulate that the level of emotional intensity he was likely to experience with the planned change in antidepressants and anticipated end of his marriage would require considerable support, whereas more exploratory therapy in the short run would only add to his emotional burden.

Similarly, with Ms K, two forms of psychotherapy were considered with different specific implications for her treatment. When both approaches were explained to the patient, she rejected exploratory psychotherapy and chose a more limited supportive psychotherapy. By being forced to consider what each psychotherapeutic approach would be likely to do in her case, I was able to present her with a meaningful choice. This is very different than a general eclectic stance saying that anyone with depression or anxiety is likely to do best with the combination of psychotherapy and medication because we should address all patients at biological, psychological, and social levels. There is a roteness and lack of granularity here that is quite unlike a POP model. The demand for specificity of proposed treatments and for postulating the nature of their effect is reinforced of course by the fact that the model will ultimately be tested by whether the predictions of the model come about or not.

7.13 A model is not a cookbook

We have been describing the basic features of POP models in psychiatry. It should also be clear from this what a POP model is not. It is clearly not a cookbook that gives the psychiatrist explicit directions for how a particular model should look for a particular patient. The generation of that model draws on the psychiatrist's expertise, capacity for pattern recognition, and imagination. The theoretical framework and training experience of the psychiatrist will inevitably influence the elements incorporated into one's models and the terms in which they are characterized and described. As discussed earlier, this need not be a problem. No one model is privileged, and multiple models are likely to yield treatment interventions that meet the patient's goals. Much more about this process will be described when we consider using the POP model as a tool for improving the clinician's skills and training in Chapter 9. I should emphasize that the purpose of constructing a model explicitly is to force the clinician to articulate the understanding to which their expertise had led them in dealing with a specific patient and provide a way of thereby predicting specific consequences of the model through which its adequacy can be assessed.

This distinction between the processes that generate the initial model and those that articulate and test it to some degree parallels a long tradition in the philosophy of science between theories of discovery and those of justification.[12] The traditional view has been that the process of discovery of a new model in science is best understood as a psychological process inside the investigator. Its mechanisms can be understood and made clearer, but they remain essentially descriptive. However, theories of justification constitute a more epistemological discipline wherein we assess the degree of confidence we should have in believing in a particular theory or model based upon some understanding of the appropriate standards of evidence. Although more recent developments in the philosophy of science have argued that the distinction between these two areas is not as crisp as originally posited,[13] distinguishing the two activities remains heuristically useful. The processes by which models are generated or 'discovered' as hypothesized patterns of propensities are considered more fully in Chapters 8 and 9. How model construction contributes to the issue of justification will be discussed when the scientific and epistemological status of model construction is considered in Chapter 10.

7.14 Model construction and data collection

Another thing that model construction is not is an instruction manual in how to collect relevant data for understanding a patient. Model construction specifically is about how the clinician reasons with the information that has been accumulated. It may indeed direct the clinician towards other areas where more data is needed. But the techniques and skill set necessary to collect information is something apart from model construction itself, although clearly a prerequisite for it. The long-standing clinical wisdom in how to conduct psychiatric interviews, how to encourage patients to share their feelings and experiences and how to help patients navigate beyond their fears and resistance to disclosure are important clinical skills that are apart from but presupposed by model construction.

However, several things about the nature of POP models are relevant to data collection in psychiatry. First, models emphasize the development of an understanding of the unique constellation of entities and interactions relevant to a specific patient. Hence limiting data collection to symptom checklists or other kinds of structured interviewing can never on its own be adequate. This is not to say that they cannot at times be useful to make sure certain areas are covered or to provide a crude screening of patients, but by themselves they can never generate the information needed to construct an individualized model.

Second, as a model is developed, predictions typically take the form of conditional hypotheses; that is, under certain types of circumstances certain types of responses by the patient are expected. This can be a guide to areas of importance in further history-taking with the patient.[14] The psychiatrist needs to look at other circumstances from the patient's past that would lead to relevant predictions that would either strengthen or weaken a proposed conditional hypothesis. In practice the cognitive process of model construction begins from the moment we start to listen to the patient's story, and is occurring concurrently with our collection of information and our efforts to establish a therapeutic alliance with the patient. Often the first tentative model is extensively borrowed from patients' own models that they present as they narrate for us their own explanations for their current circumstances.

7.15 The role of diagnosis in model construction

The consideration as to how various levels of explanation are integrated in model construction leads to the question of the role of a diagnosis. After all, in the current DSM (*Diagnostic and Statistical Manual)*, a diagnosis is specifically an atheoretical construct. So, on what level of explanation does it live? How useful is it in constructing a model of a specific patient? I would argue that a diagnosis is essentially a placeholder that can be used to occupy a piece of a POP model that stands-in for some of the relevant but poorly understood propensities and patterns of influence that we are positing in that patient. It allows us to bring to bear certain empirical information and generalizations, such as those from clinical trials or naturalistic studies, that may suggest aspects of our understanding or predictions of useful interventions, even when we in fact have only a very limited knowledge of the processes involved.

How big a role that placeholder plays in the model will vary from patient to patient. In the case of Ms K, I postulated that some disposition to anxiety that could be captured by the label 'generalized anxiety disorder' (GAD) was present in Ms K. However, we know very little about what underlies GAD. In this model, the diagnosis both serves to remind one of the potential relevance of familial factors as well as the possible relevance of certain standardized treatment options that have been shown of value in groups of patients with GAD. Familial factors are likely to include genetic as well as environmental factors. In the case of Ms K, however, at least some of the important psychological factors are quite understandable from her early developmental and family history. Hence one can build a psychological level of explanation into the model quite easily without relying on the diagnosis of GAD.

In contrast, we have little to go on in understanding any specific biologic mechanisms or genetic factors that may be specific to Ms K. However, by subsuming her under the class of patients with GAD, diagnosis-derived information on the genetics of GAD can become part of our hypothesized propensities in our specific model for Ms K. Furthermore, the clinician wants to be able to use the empirical data that is available suggesting that many patients who meet criteria for GAD benefit from selective serotonin reuptake inhibitor antidepressants (SSRIs). Without having a clear understanding of why that is true, in either psychological or biological terms, the pharmacologic treatment can only be built into the model by positing the presence of some unknown mechanisms that underlies the category of GAD. I should note that in this case the GAD diagnosis is employed even though technically she does not meet the criteria for the disorder by current DSM standards. I hypothesized that she had enough features of GAD that she likely shared some of the mechanisms present in that condition so that she was in turn likely to benefit from treatments found effective in many patients with that diagnosis.

This issue can also be illustrated in the case of Mr Z. Again, several factors best characterized in psychological terms can be identified as contributing to his propensity to depression and can be extracted from a good developmental and family history. The diagnosis of major depressive disorder can be used to fill in for our fundamental ignorance about the neuromolecular mechanisms involved, and how antidepressant medications in fact work. Applying the category major depressive disorder allows us to posit some ill-defined mechanisms likely shared with other patients meeting criteria for major depression, and to draw upon the empirical clinical trial data that suggests that antidepressant medications have a good chance of efficacy for these conditions.

This use of diagnostic categories keeps their role in some perspective. They do not become the defining feature of any patient's condition and in many cases may be a relatively minor part of the total model. Some proposed interventions will work through that placeholder. Many other aspects of our understanding of the case and of the role of specific treatments can be best understood by looking at other elements in the model that have little to do with the assignment of a diagnosis.

This circumscribed use of diagnoses in understanding our patients can be appreciated by contrasting it with the approach of the expert in affective disorders with whom Ms M worked during one phase of her treatment. For that expert the diagnosis essentially was the model. He sought to explain the nature of her difficulties almost entirely as an instance of a particular disorder. Everything was explained in terms of that disorder, even her striking pattern of emotional responses to relationships beginning and ending was reduced to a set of nonspecific stressors that served as triggers for the cycling of her illness. I cannot help but wonder if this psychiatrist was prone to this sort of thinking precisely because his claim to fame was being viewed as an expert in a condition, namely bipolar disorder. Perhaps he saw his expertise as consisting in expertise about the disorder not in expertise about an individual patient.

As another example of a case where the patient's DSM-based diagnosis played a minor role in the development of a POP model and implementing interventions based of it, consider Ms J.

Ms J, a middle-aged woman, described a persistent dysthymia since early adolescence combined with extremely low self-esteem and constant self-criticism. She had a tumultuous adolescence with high risk-taking, sexual promiscuity, and intermittent drug abuse, although she never developed a persistent addictive pattern of substance use. After adolescence she abandoned her impulsive behaviour and became quite conventional. She completed college and had a successful career in a middle-management position. She married in her late 30s to a man who treats her well and clearly cares about her. She chose to have no children. Throughout her adult life she continues to berate herself for her behavioural excesses in adolescence and feels that she interacts so poorly with others that eventually everyone comes to dislike her. There is in fact no objective evidence for this. She describes no more than the usual amount of conflict with coworkers over the years, and has friends despite her tendency to pull away from them in anticipation of rejection.

Ms J grew up in a small Midwestern city that was quite conservative and had a small collection of socially elite families who had been in the city for multiple generations. These families tended to socialize primarily with one another and how they were viewed by the other families in their social tier was seen as very important. Ms J's father's family was among this group. Her mother's family was not and her father was seen as 'marrying down'. Furthermore, her mother was extremely emotionally labile and erratic. She would behave quite normally for periods of time but then become overly loquacious and quite rude, and at times paranoid. She became something of a scandalous character within the social circle of elite families. All this was profoundly embarrassing to the patient. Furthermore, when Ms J began to act out in adolescence, in contrast to her very proper and well-behaved brothers, people frequently commented that she was likely to turn out just like her mother. She very much took this to heart and saw her own behaviour as a source of embarrassment and shame just like her mother's. She thought that no matter how much she worked to repress it throughout her adult life, it would be only a matter of time before her mother's traits would show in her own behaviour causing embarrassment to herself and driving others away. Over the years Ms J had sought psychiatric help on multiple occasions and both psychotherapy and pharmacotherapy were employed. Antidepressants and mood stabilizers had very minimal effect on her dysthymia. Several courses of psychotherapy had little impact.

What would be the role of a diagnosis for this patient in a useful POP model? Her mother in all probability had some sort of bipolar disorder. Outside of her erratic and impulsive behaviour in adolescence, Ms J's subsequent life shows no suggestion of a bipolar process. She certainly meets the diagnostic criteria for chronic dysthymia, but what role does that diagnosis play in developing a model? Had they not already been tried, the dysthymic diagnosis could support antidepressant treatment trials. But she had already failed trials from every major category of antidepressants.

I would suggest in this case that the most appropriate and useful model would see her unyielding dysthymia and self-loathing as resulting from an upbringing characterized by an extremely shame-oriented micro-culture, a mother's behaviour that induced extraordinary embarrassment in that culture, and a socially reinforced pathological identification on the part of the patient with her mother. Ms J's official diagnosis based on her symptoms had little role to play here.

7.16 Models are bidirectional: They explain and predict

Another feature of POP models is their bidirectionality. A model is generated in the process of an encounter between psychiatrist and patient that occurs at some point in time in the trajectory of the patient's life. We can use the model to look backwards to explain what led to the patient's current circumstances. Confirmatory evidence for this use of the model typically comes from an ongoing process of detailed history-taking. New information that emerges over time that was unknown to us in postulating the original model serves much like a prediction since it played no role in generating the model itself. The new data that emerges in the ongoing treatment relationship can inform the model for us even though the revealed information occurred in the past.

Alternatively, the model can begin with the patient's current condition and be used to make predictions about the future in terms of how the patient is likely to respond to certain environmental perturbations. An important category of such perturbations is of course the treatment interventions provided by the psychiatrist. Here future responses are the predictions by which the adequacy of the model is judged. These two uses of the model reinforce one another, and in some cases blend in striking ways. In many forms of psychodynamic psychotherapy, for instance, the exploration of the past and the elaboration of the model in those terms serves not only to explain how the patient got to their present circumstance but the process itself is expected to generate clinical improvement in the future. Thus, it also is relevant to future oriented prediction.

Thus, for Ms M, I hypothesized a pattern of early family relations that explained her repeating patterns of unsuccessful relationships with men and the dramatic alternating mood states that accompanied them. This is using the past to explain how the patient arrived at her current situation. Using that understanding, I then looked forward to make predictions about both her future relationships if nothing changed, as well as to postulate interventions that could ameliorate that pattern. Furthermore, the key intervention, psychodynamic psychotherapy, involved bringing the patient to an understanding and acceptance of that historical explanation in the belief that this very process would allow her to overcome that destructive pattern in the future.

7.17 The importance of pruning in model construction

I have emphasized the important role for disciplined and informed imagination in generating a POP model. I also pointed out that different clinicians, with different training backgrounds and levels of experience are likely to generate different models for the same patient. I also discussed why this need not be a problem, as more than one model is likely to be adequate to achieve the patient's goals. However, it is essential to also remember that this is not an 'anything goes' process. There are multiple constraints that require the psychiatrist to prune away many candidates from the range of potential models.

Epistemological pruning requires that our model be compatible with existing knowledge in the field relevant to our patient. Some pruning is pragmatic, for instance when certain possible treatments are simply not available for the patient in a given circumstance. There is also negotiated pruning, when certain models and the interventions derived from them do not match the patient's understanding of her own situation or values, and hence are unacceptable to the patient. The case of Mr D is a particularly clear instance of this form of pruning.

Finally is what can best be characterized as ethical pruning. Our obligation as physicians to first do no harm requires that we conscientiously consider the potential adverse effects that could come from any intervention we are proposing. This can take two general forms. The first is to consider and discuss with our patients the potential side effects and other adverse events that can follow from the treatment even if our model is correct. This is clearer with biologic interventions but can apply to psychological and social interventions as well. The second form is the possibility that some aspect of our model is wrong, and that our proposed intervention will fail to bring about the desired results or may even cause unintended, harmful consequences. This requires another exercise of the clinician's critical imagination. One must ask oneself to actively consider what might go wrong, to discuss those possibilities with the patient and together decide whether the risk is acceptable and in what ways it might be minimized. These constraints apply both to the generation of the initial model for the patient as well as for any subsequent required revisions.

I have elsewhere argued[15] that the process of POP model construction encourages a healthy humility in the psychiatrist that itself constitutes a clinical virtue and an ethical good. Both the clinician's understanding of the patient's difficulties and the generation of options for intervention only make sense in terms of the unique goals that the patient brings to the clinical encounter. The model develops through a process of exchange between psychiatrist and patient leading to some shared perspective that becomes the basis for the clinical work. Furthermore, over time changes in the patient's goals and circumstances often require a frequent renegotiation of the model and the treatment decisions flowing from it. Of course, the success or failure of interventions derived from the model may also require further modifications or even the need for a new model.

This contrasts sharply with the psychiatrist's stance in traditional EBM. Here it is assumed a conscientious assessment of the existing evidence-based literature yields our current best judgement as to what is the optimal intervention for a group of patients of which our patient is a member. While the clinician acknowledges the patient's right to reject such a treatment recommendation, it is hard to resist the assumption that the psychiatrist already knows the 'right answer' and that one's job is to persuade the patient to accept this.

A clear example of this is the case of Mr D, where prior psychiatrists so strongly pushed the need for continuous antipsychotic medication that he repeatedly left treatment. How much of what we describe as patient non-adherence springs from this sort of subtle or not-so-subtle arrogant and parental attitude? By developing a model based on a shared understanding of intervention strategies that made sense in terms of his goals, Mr D was able to establish a constructive therapeutic alliance that allowed him to achieve sustained stability over many years.

7.18 When models need to be explicit

I have provided a general characterization of what a POP model looks like, what are its salient elements and features, how it serves to represent the product of a psychiatrist's reasoning about a specific patient, and how it leads to potential interventions. As previously stated, model construction need not be a conscious process, especially for the experienced clinician. Many of the patients we treat bring concerns that are sufficiently straightforward and that fit known patterns with sufficient accuracy that an experienced clinician can simply recognize the pattern at work and suggest appropriate treatments. However, an explicit understanding of models should always permit the psychiatrist, if called upon, to articulate a POP model that reflects one's understanding of that individual patient.

Several circumstances may demand making models explicit. One is in presenting a case to another involved clinician who may have a different perspective. Other instances might be when anomalies in the patient's experience and behaviour emerge that violate typical patterns, or when treatment decisions are simply not working. In section 9.9, I will argue that making models explicit is also a very useful training technique with novice clinicians.

While it may appear cumbersome to construct an explicit model even in a very standard case, it may very well help a novice clinician to develop the thought processes necessary for good clinical work. In this sense it is not unlike the medical student who is asked to learn how to conduct a full history and physical examination that typically takes multiple hours to complete. Learning how to do this is essential to the clinician's training and medical understanding, but once one becomes an established clinician, one is unlikely to ever do that kind of complete physical examination again, but rather will draw on the parts of it that are appropriate to a specific patient and circumstance. But the template of that complete examination remains in the physician's head, suggesting the range of possible areas to be explored when anomalies emerge in the treatment of a given patient. In the next two chapters we will engage in a more detailed discussion of the practical uses of model construction in psychiatry, that is looking at models as pedagogical tools.

Endnotes

1. Klein 1998, p. 49
2. Gigerenzer 2008
3. Klein 1998, p. 34
4. Munson 1981
5. Cartwright 1999. pp. 49–53
6. Cartwright 1999, p. 74
7. See Chapter 3, endnote 3
8. Jaspers 1997
9. Dutton and Aron 1974
10. Ghaemi 2003
11. Engel 1977

12. Reichenbach 1938
13. Giere 1999
14. This movement back and forth between very specific information about an individual patient and more general considerations is a process that occurs in many aspects of POP model construction. The psychiatrist is expected to bring to bear their general knowledge base about groups of patients or basic science mechanisms and apply it to the individual patient. Conversely, the careful consideration of the individual patient provides a more nuanced understanding of that general knowledge. A branch of epistemology termed hermeneutics, which focuses on all knowledge as a form of interpretation, discusses this back-and-forth process in considerable detail as the 'hermeneutic circle' (see Phillips, 1996). The complicated relationship between hermeneutics and the POP modelling approach is considered more fully in section 10.2.1
15. Heinrichs 2015

Chapter 8
How we reason and make decisions

8.1 Models as deliberation

In Chapter 7 we stepped back from the series of clinical case studies to provide a more general description of pattern of propensities (POP) models. We offered it as an explicit articulation of the often-tacit method of deliberation typical of the experienced psychiatrist engaged in the activity of psychiatry proper, that is treating an individual patient. A POP model consists of propensities and the paths of influence or interaction between them. In this and the next chapter, we will consider the value of POP models as tools to improve clinical decision-making and to contribute to the initial and ongoing education of psychiatrists. Education here refers both to developing the competence of trainees and novices in the field, such as psychiatric residents, to turn them into experts, and to enhancing the performance of established and experienced psychiatrists.

I should be clear about what aspect of education and training we are referring to in this context. Developing a knowledge base for the 'informing disciplines' of psychiatry is not in question. The psychiatrist needs a strong knowledge base in psychopathology, clinical therapeutics, and a range of other biologic and psychosocial disciplines that have relevance to understanding psychiatric patients. The psychiatrist also needs to develop a skill set for establishing rapport with patients by learning to develop and utilize their capacity for empathy, and to acquire the skills of eliciting relevant information in a clinical interview and a mental status examination. None of these things are the focus here. Rather we want to consider using the POP model as a tool to help psychiatrists improve their ability to reason about the mass of information obtained about individual patients to generate an understanding of their difficulties and to suggest potential interventions to ameliorate their distress.

In short, we are talking about inculcating an effective method of deliberation about the individual patient to facilitate good clinical decisions regarding the treatment of that patient. This includes both the formulation of an initial treatment plan and the revision of that plan in the light of

How Psychiatrists Make Decisions. Douglas W. Heinrichs, Oxford University Press. © Oxford University Press (2025).
DOI: 10.1093/9780198960720.003.0008

the patient's subsequent treatment experience and circumstances—reasoning in the service of decision-making. Before addressing this issue in reference to psychiatry specifically, it is helpful to consider human reasoning and decision-making in general. Our current understanding is the product of centuries of self-reflection by members of our species. This will be the focus of the current chapter. I urge readers to be patient with this detour, as it will shed valuable light on our understanding of psychiatric deliberation, which will be taken up in the next chapter.

8.2 Deduction in human reasoning

Since at least the ancient Greeks, humans have attempted to use their capacity to reason to reflect upon the reasoning process itself. There is a long and fascinating history to this quest. Its legacy forms an important piece of the subject matter of philosophy and psychology. The nature of reasoning is tied up with the perennial human desire for certainty in our understanding of the world and ourselves. How do we know things and how confident can we be in what we know? While volumes have been written on this topic, here we will highlight several features of this history leading up to current concepts about human reasoning and decision-making that are highly relevant to our understanding of how clinicians reason and how that reasoning can be enhanced.

Since at least the time of Aristotle, a distinction was made between what in today's language we call deductive versus inductive reasoning. Deductive reasoning refers to a set of logical processes by which one can move from certain premises that are taken to be true to reach certain conclusions that can be justifiably asserted to be true as well. The classic paradigm for this kind of thinking is the syllogism. It typically involves a statement of a generalization which is taken to be universally true, referred to as the major premise, combined with a particular statement about an individual, referred to as the minor premise, which is also taken to be known as true. The logical deductive process involves applying correct rules that allow one to draw a further specific conclusion about the individual case in question. Take for instance the well-worn classic example:

> All men are mortal. (Major premise)
> Socrates is a man. (Minor premise)
> Socrates is mortal. (Conclusion)

Note that the conclusion is reached without acquiring any new information. It simply follows by the rules of logic from the information contained in the two premises. An assumption of deductive reasoning is that the rules for this kind of reasoning are infallible; that is if the premises are correct, the conclusion must be correct unless one has committed a logical error. The Greeks had, and have left as an ongoing legacy in Western thought, a love affair with the notion of certainty in human knowledge. Hence deductive reasoning was seen as the pinnacle of human thought, and true knowledge could only be had in disciplines that employed deductive reasoning. Such disciplines were characterized as *episteme*. Disciplines where such deductive thinking was impossible yielded only beliefs or opinions, which may have a high likelihood of being true, but could not be known with certainty. Such disciplines, that included many practical skills, were generally characterized as *technai*.

The paradigm case of *episteme* rooted in deductive reasoning is mathematics, especially geometry. A few basic axioms, taken to be self-evident and universally true to the Greeks if not to modern thinkers, are used to generate a range of knowledge statements purely by constructing chains of logical deductions. For instance, from the definition of a triangle I can deduce that for any triangle the sum of the angles is 180 degrees. I do not need to examine a large number of triangles to verify this generalization. Furthermore, if I know that shape X is a triangle, I also know that the sum of its angles is 180 degrees without empirically measuring them.

This sort of reasoning is all fine and good in a universe of abstract thought. However, how far will deductive reasoning, with its bragging rights about certainty, get us in any discipline intending to provide knowledge about the real world, even in highly mathematized fields such as physics? Where do the generalizations—the laws of nature, if you will—come from in such disciplines? They clearly are not self-evident. Typically, they result from repeated observations that show us a consistent pattern. The law of gravity and the value of the gravitational constant is not self-evident in its details, but was derived from repeated observation of the interactions of bodies that came to be understood in terms of their masses and distances from each other.

8.3 Induction in human reasoning

These latter practices required a second form of reasoning known as induction. The goal of induction is to derive a valid generalization from the observation of multiple instances of a class of events. The assumption is that the more instances one observes, the greater confidence one can have in the resulting generalization, assuming all the instances yield the same result. Once such a consistent generalization is found, it can serve as the major premise for further deductive thinking. The inclusion of induction allows human reasoning to extend to the vast array of real-world situations, but at a cost. That cost is the loss of certainty. Unless we know we have observed every member of a class, it is possible that if one continued to collect instances one would find exceptions to the generalization. Consider the following syllogism:

All mammals produce their young by live birth. (Major premise)
X is a mammal. (Minor premise)
X is the product of a live birth. (Conclusion)

For a long time, this would have been viewed as a valid piece of reasoning. But with the discovery that the platypus and four species of echidnas are mammalian species that lay eggs, this bit of reasoning is clearly not valid. Now some valid reasoning can be salvaged from this sort of situation, but only by accepting the loss of certainty that comes in making the generalization a probabilistic statement. Hence, we can say the following:

Most mammals produce their young by live birth. (Major premise)
X is a mammal. (Minor premise)
X is probably the product of a live birth. (Conclusion)

8.4 The role of statistics and probability

This bit of reasoning seems valid as far as it goes. But exactly how confident should we be that X is the product of a life birth? The effort to answer this sort of question with some precision can be seen as the driving force behind the development of much that we have come to view as scientific method as well as the application of statistics and probability in modern inquiry. It is interesting that these methodologies are relatively recent creations of the human mind, specifically having their origins in the seventeenth century.[1] In a very real sense, the passion for certainty has not been abandoned but shifted: If we cannot be certain about our conclusions, perhaps we can be certain about how uncertain we are. Thus, to the modern mind it would be extremely natural to reason that if we know the total number of platypuses and echidnas in the world and the total number of mammals, we can construct a fraction (call it Px) that represents the probability that X is the product of a life birth.

But there are limitations here. The number of platypuses and echidnas would of necessity be estimates derived from samples and hence not precisely accurate. Considerable effort in the development of statistics has been directed to characterizing the magnitude of such potential sampling errors. Furthermore, the actual value of Px is tremendously influenced by how we selected the X of interest in the first place. If I live in North America and select the first mammal I encounter, Px would be infinitesimally large. If I do the same thing but live in Australia, it would be somewhat lower. If I work at a centre dedicated to the preservation of the platypus, Px would be quite low. To address these kinds of issues and others like it, extensive methodological procedures have been developed as paradigms for conducting research in the biological and social sciences. The hope is that a meaningfully precise Px can be calculated when a standardized methodology is employed.

The importance of statistics and probability and its application in scientific method is extremely important in modern science, and the full range of its implications are beyond the scope of our discussion. For present purposes, I want to make two points. This probabilistic generalization, that most mammals are the product of live births, can be used to generate two sorts of conclusions. On the one hand, one can select a specific mammal X and draw a conclusion about the likelihood of it being the product of a life birth. However, one could also dispense with having a specific subject for the minor premise and draw the conclusion that the next mammal I encounter is likely to be the product of a life birth. At first glance these may appear like similar situations. But in real-world terms there is a profound difference. In the latter case we know nothing about the next mammal we will encounter. Hence the tools of probability and statistics do in fact usually yield the best estimate of Px. But in the situation where we have a specific X in hand, the same procedures may not give us the best available estimate. If I have X in hand, I might look and say, 'This is a dog, not a platypus or echidna, and therefore I know that X is the product of life birth'. From the point of view of mathematical probability and statistics, bringing to bear that specific knowledge about X constitutes a form of cheating. To correctly apply such methodology requires that we remain ignorant about many specific features of X. The statistician Herbert Weisberg[2] argues that in practice much valuable information is lost in areas like modern medicine with the thoughtless application of such 'wilful ignorance' in an attempt to be scientifically rigorous. For the practising psychiatrist making decisions about a specific patient, the situation is much more like having an X in hand.

The second point is that in our modern intellectual culture, the principles of modern statistical thinking and 'scientific method' are not simply viewed as the appropriate methods for specialized areas of inquiry, but are often seen as constituting the standards for judging the validity of all rational human thought. Optimal human reasoning should follow the rules of logic, both inductive and deductive, as well as the more recently characterized logical principles of probability and statistics. As we shall see, this assumption is important as we turn more directly to a consideration of a modern understanding of human decision-making.

8.5 The concept of the 'rational man'

While humanity's attempt to understand the framework and standards of logic and human rationality go back thousands of years,[3] serious study of human decision-making in concrete situations is a much more recent phenomenon.[4] It is true that Aristotle, while viewing *episteme* as the most perfect form of human knowledge, acknowledged that many forms of human intellectual activity do not lend themselves to this kind of reasoning with its resultant sense of certainty. Aristotle included the *techne* of medicine explicitly in this latter category. Aristotle had a considerable amount to say about medicine as a discipline,[5] not surprising given that his father was an esteemed physician. He spoke of a particular intellectual virtue, *phronesis* or practical wisdom[6], that an individual needs to effectively make decisions about concrete and specific situations given their intrinsic uncertainty. But what he had to say about this virtue remained rather vague and poorly characterized, and he gave very little concrete guidance in understanding the process of human decision-making. The serious, empirical study of human decision-making only gained momentum in the late twentieth century. Much of this work was driven by very practical concerns, hence it has been produced not just by cognitive psychologists but often by behavioural economists and other practical thinkers in the world of business, leading to several Nobel prizes in economics. Several overlapping schools of thought have emerged over time in this area, again originally applied primarily in economics.

The first and oldest was based on the concept of the 'rational man'. The notion here is that useful economic theories and predictions could be generated by assuming that individuals make decisions applying rational assessments of their self-interest as defined by the rules of logic and probability (expected utility theory).[7] Over time it became increasingly clear that models of economic activity employing the concept of the rational man generally did a very poor job in predicting the outcome of that activity. It appears that humans simply do not apply the rules of logic and probability accurately in making economic and other important practical decisions in their daily life. This led to a certain crisis in economic thinking. If humans were to some considerable degree irrational in their decision-making, how could one possibly make effective models to predict their behaviour?

8.6 Two-systems theory and cognitive biases

A response to this challenge led to what can be seen as a second phase of work in understanding human decision-making. The assumption here is that human cognitive capacity was not generated in a vacuum as some sort of abstract reasoning device constructed according to the rules of logic and probability theory. Rather it is the product of cognitive evolution in a species seeking to optimize its survival and flourishing in the real world. Effective cognitive strategies could persevere even when violating the laws of formal logic and probability, so long as they were an aid to the organism's survival. If this is the case, it should be possible to identify consistent patterns of technically illogical reasoning and inference in the human thinker. If humans possess a pattern of irrationality that follows certain definable patterns, perhaps their behaviour would become predictable after all. This approach led to a fascinating body of experimental efforts best embodied in the work of Daniel Kahneman and Amos Tversky, leading to the award of the Nobel Prize in economics to Kahneman. An excellent explication of the theory resulting from this work, as well as a detailed consideration of some of the empirical studies on which it is based, can be found in Kahneman's 2011 book, *Thinking, Fast and Slow*.[8]

The fundamental construct here is that there are two independent cognitive systems that humans have evolved, termed 'System 1' and 'System 2'. System 1 is from an evolutionary point of view more primitive. It overlaps extensively with systems found in the cognition of other mammalian species. Confronted with a situation demanding a decision, System 1 works quickly and intuitively to generate essentially a gut level response as to what to do. That judgement is seldom based on rules of logic or probability in any formal way. Rather it is the instinctual activation of a particular bias in situations of a certain sort that from an evolutionary perspective has been associated historically with enhanced survival value. System 1 has its virtues. It can be deployed quickly, thereby allowing a rapid response to a critical situation. It also involves minimal expenditure of cognitive or emotional energy, and therefore can be utilized even when the individual is exhausted, distracted, or in a state of deprivation such as hunger and thirst. The limitation of System 1, of course, is that many of the complex decisions that people must make in the modern world are quite different from the more primitive environmental demands that this system evolved to address.

Hence System 1's actions are likely to lead to many poor decisions. Note that these errors in judgement will not be random. They follow certain characteristic patterns based upon the evolved propensities of System 1. These characteristic patterns of error are collectively referred to as 'cognitive biases' or 'cognitive illusions'.[9] Much of the empirical work in this field involves using highly contrived situations in laboratory settings in which an individual must make a judgement. A range of consistent patterns of error emergence from this work, leading to a list of common cognitive biases.

System 2 is a later evolutionary acquisition. It is a higher-level system that allows the individual to step back from the situation and conduct a dispassionate analysis, based on the application of rational standards of logic and probabilistic reasoning. System 2, in essence, provides a mechanism that can allow the individual to correct initial and often erroneous gut level responses, and thus reach a more rationally justifiable decision. This gives some hope for our hapless human thinker, but there are numerous obstacles to the effective application of System 2. First, it takes time. Compared to System 1, System 2 is slow and labour-intensive. This makes it difficult to employ in situations demanding rapid action. Even with time for a careful analysis, the operations of System 2 demand a great deal of psychic energy and sufficient calm for a detached consideration of the issue at hand. The assumption is that humans are essentially cognitively lazy, and have a natural reluctance to apply System 2.[10] Furthermore, the effective operation of the system is impeded if the individual is fatigued, in a state where basic needs have not been met such as hunger and thirst, and where there is a highly charged emotional valence to the situation.[11] As many important human decisions must be made in a context where one or several of these factors are operative, these limitations on the effective operation of System 2 are severe indeed. So, despite our best intentions, we are destined to frequently fall victim to the cognitive biases of System 1 as we navigate our way through life.

8.7 Some examples

A consideration of a few examples of cognitive biases and the kind of experimental setting in which they are elicited will be helpful to get a concrete feel for this theory.

A group of undergraduates were presented with the following problem:

> Linda is 31 years old, single, outspoken, and very bright. She majored in philosophy. As a student she was deeply concerned with issues of discrimination and social justice and also participated in antinuclear demonstrations. Which of the two alternatives is more probable:
>
> Linda is a bank teller,
>
> Linda is a bank teller and is active in the feminist movement?[12]

About 85% of the undergraduates chose the second alternative. This was interpreted as representing an instance of the 'conjunction fallacy' and is a clear violation of the rules of formal logic. The probability of being a member of class A must be equal to or greater than the probability of being a member of both classes A and B. When presented in this abstract form almost everyone can see its validity. Hence the chance that Linda is a member of the class of bank tellers cannot be less probable than the dual condition of her being both a member of the class of bank tellers and the class of activists in the feminist movement. Kahneman argues that incorporating the added details about Linda make the option of her being a feminist bank teller more persuasive and plausible to System 1, by providing an intuitively appealing causal pathway from her earlier activities to her later ones. This intuitive reaction overrides System 2's logical calculation of probability, creating a 'conjunction fallacy'.

Or consider the following problem also presented to a group of undergraduates:

> A certain town is served by two hospitals. In the larger hospital about 45 babies are born each day, and in the smaller hospital about 15 are born each day. As you know about 50% of all babies are boys. However, the exact percentage varies from day to day. Sometimes it may be higher than 50%, sometimes lower. For a period of one year each hospital recorded the days in which more than 60% of the babies born were boys. Which hospital do you think recorded more such days?[13]

Slightly over half of the undergraduates judged that it would be the same for both hospitals. The rest were evenly split between the larger or the smaller hospital. The correct answer is the smaller hospital. An established principle of probability theory states that variability is greater in smaller samples than in larger ones. Hence the frequency with which the sample will exceed 60% boys will be greater for the smaller hospital. Moreover, smaller hospital will have more days in which more than 60% of the babies were girls as well. Here the bias of insensitivity to sample size resulted in most undergraduates getting this problem wrong. Kahneman stresses that System 1 automatically looks for and accords great weight to a coherent story and relevant causal factors to make predictions. In this case where there is no apparent causal reason for a gender difference in births between the two hospitals, the resulting tendency to assume no difference overrides the consideration of the higher variability of small samples over large ones that would enter the considerations of System 2.[14]

The tendency for plausible causality to override statistical likelihood can be illustrated with a psychiatric example. A young adult presents with an episode of major depression. She has a strong family history of mood disorders, including multiple relatives with bipolar disorder. It is easy for the clinician in this case to assume it is probable that this patient will go on to develop a bipolar disorder over time. After all we know that there is genetic causality in bipolar disorder. However, to make an accurate prediction, we need to remember that unipolar depression is considerably more common than bipolar depression. Therefore, statistically there is a greater chance that our patient will have a unipolar than a bipolar illness, despite her family history.

8.8 The rational man is still the ideal in two-system theory

There are two features of this two-system theory of human cognition that should be noted for our purposes. The first is that while the theory clearly argues that most humans do not reason in a truly rational manner most of the time, the rules of logic and the principles of probabilistic thinking are still seen as the appropriate standards for judging valid rational thought. System 1 operates much of the time and is frequently in error, but System 2, in those situations where it can operate optimally, yields 'correct' thinking by applying norms of logic and probability. Thus, while it yields many different practical implications for predicting human behaviour, the two-system theory is epistemologically quite consistent with the theory of the rational man. For both, good reasoning and decision-making should follow the rules of logic and probability.

A corollary of this position is that for any problem there is either a literally correct answer if all needed information is available, or an optimal answer that has the greatest probability of being correct when full information is not available. So, the goal is assumed to be to find the very best answer to a question or to make the very best decision in a situation where action is called for. Decision strategies based upon this way of thinking are often called optimization or maximizing strategies.

The second observation is to note the general way information is collected in studies of cognitive biases. The experimental subject, usually the ubiquitous college undergraduate, is presented with a set of statements from which one is expected to draw the correct conclusion. These statements are generally provided without a larger contextual framework to provide any situated understanding of the setting in which the judgement must be made. If one is only looking to see if the rules of formal logic and probability theory are being applied, this lack of context is understandable. Real-life human decision points, however, seldom occur this way. They are typically embedded in an elaborate context that the individual uses to make sense of the specific judgement one is called upon to make. Furthermore, the subjects in these experiments are essentially an 'everyman' in that they are not thought to bring any special expertise to bear relating to the subject matter or content of the judgement being made. As we will see, the processes that human decision-making employs are heavily influenced by the level of expertise the individual has regarding the subject matter in question.

The advocates of the two-system theory of human cognition do argue that making an individual consciously aware of the range of cognitive biases to which one is subject, as well as the kinds of conditions that impede the operation of System 2, can lead to some success in avoiding or mitigating those biases in the service of more rational thought. Nonetheless, the power of System 1, and the many factors that can interfere with the effective operation of System 2, suggests that despite one's best efforts, biased and irrational decision-making is often inevitable. Thus, there is a fundamental pessimism about human reasoning at the heart of this theory.

8.9 Heuristics in human decision-making

One response to this pessimism has led to a third wave of theorizing about human rationality and decision-making. This is sometimes called the heuristic approach or the theory of ecological rationality. This position accepts that much good human reasoning is based on evolved propensities to apply certain rules of thumb in making decisions about certain situations the individual confronts. These often do not follow the rules of formal logic and probability. Indeed, the application of these rules of thumb, or heuristics, can at times lead to error. However, this school of thought emphasizes that when applied in an appropriate environmental context, a heuristic not only often yields a good solution, but frequently is superior to an attempt to apply rules of formal logic and probability. The goal is to know the right environmental context to use a given heuristic. The notion of ecological rationality implies that good decisions and judgements result from the correct match of a heuristic strategy and an appropriate environmental context for applying that strategy.

One of the leading workers in this field is Gerd Gigerenzer, the director of the Center for Adaptive Behavior and Cognition at the Max Planck Institute. The following discussion of this approach, including several of the examples, are drawn from his excellent book, *Rationality for Mortals: How People Cope with Uncertainty*.[15] He characterizes the study of heuristics in the following way:

> The science of heuristics has three goals . . . [first] to analyze the adaptive toolbox, that is, the heuristics, their building blocks, and the evolved capacities exploited by the building blocks . . . The [second] goal is to determine the environmental structures in which a given heuristic is successful, that is, the match between mind and environment (physical and social). This analysis includes the co-evolution between heuristics and environments . . . The [third] goal is to use the results of the study of the adaptive toolbox and [the match between mind and environment] . . . to design heuristics and/or environments for improving decision-making in applied fields such as healthcare, law, and management.[16]

The way a heuristic strategy is built up from more basic evolved capacities is illustrated by Gigerenzer[17] by considering a very unconscious action embedded in a skilled response. How does a baseball outfielder know where to run to catch a fly ball? In general, a fly ball follows the path of a parabola. In the modern world we have differential equations that allow us to calculate the path of a parabola and hence locate where the ball would land. I suspect that most ballplayers have little knowledge of differential equations. Furthermore, they do not have many of the pieces of data needed to make an accurate calculation, i.e. the ball's initial distance, initial velocity, projection angle, air resistance, wind, and spin. It seems preposterous to posit that humans in their course of evolution generated an unconscious capacity to make such calculations. Furthermore, there is no time for this, as the outfielder must respond at once.

Fortunately for the game of baseball and for a range of other human activities, the 'gaze heuristic' deals with this problem: Fixate your gaze on the ball, start running, and adjust your speed so that the angle of gaze remains constant. This heuristic draws on a very basic evolved capacity. Indeed, if one follows the path of a dog running to catch a frisbee, it appears that this same strategy is being employed. This heuristic is characterized as being fast, frugal, transparent, and robust. The gaze heuristic is fast in that it allows the player to respond almost instantly. It is frugal in that it requires very little information, just the angle of gaze. There is no need for all the data that would be required to apply differential equations. It is transparent in that it can be easily understood and taught to a novice. It is robust in that the ability can be applied to a range of new situations.

Using an understanding of evolved human capacities to develop useful heuristics can be seen in many contexts. Let us consider an example directly relevant to medicine. Forty-eight physicians, with an average of 14 years of experience, were asked to draw a conclusion regarding risk prediction in breast cancer. The same information was given to them all, but to half it was presented in terms of conditional probabilities and to the other half as natural frequencies as follows:

> **Conditional Probabilities:** The probability that a woman has breast cancer is 0.8%. If she has breast cancer, the probability that a... mammogram will show a positive result is 90%. If a woman does not have breast cancer, the probability of a positive result is 7%. Take, for example, a woman who has a positive result. What is the probability that she actually has breast cancer?
>
> **Natural Frequencies:** Eight out of every 1,000 women will have breast cancer. Of these eight women with breast cancer, seven will have a positive result on mammography. Of the remaining 992 women who do not have breast cancer, some 70 will still have a positive mammogram. Take, for example, a sample of women who have positive mammograms. How many of these women actually have breast cancer?[18]

The correct answer is 7 out of 77, or about 9%. Strikingly, of the 24 physicians given the data as conditional probabilities, only two got the correct answer with two-thirds of them giving answers above 45%. For the 24 physicians given the data as natural frequencies, 11 answered correctly and all but five answered under 15%. In discussing this and similar studies, Gigerenzer points out that it is quite natural for humans to think in terms of frequencies whereas statistical and probabilistic reasoning is a recent human invention dating from the seventeenth century, and must be explicitly learned. While more rigorous training in statistical inference and probability for physicians is desirable, it is easier to improve data interpretation by physicians by presenting the data as natural frequencies whenever possible.[19]

A heuristic, when applied in a real-life setting to which it is well matched, can yield a superior result when compared to a formally logical or calculated response. This is especially true when one goes out of contrived experimental settings and considers complex real-life decisions. Gigerenzer[20] notes that contrived laboratory tasks often make formally calculated responses feasible by editing a real-world problem to make it accessible to some optimization calculus by strategies such as limiting the number of competitive responses and choice alternatives, by providing quantitative probabilities, or by ignoring complex environmental constraints. He notes that in the real world most situations calling for a decision are sufficiently complex that an optimizing strategy is often unknowable or computationally intractable; that is, no mind or machine can find the optimal solution in a reasonable time.

He goes on to argue that many of the alleged errors in reasoning deriving from cognitive biases that have been identified by the two-system school of research derive from the artificiality of the setting. Consider the example of Linda the bank teller that was presented earlier. Abstracted from all context it is indeed logically correct that Linda is more likely to be a bank teller than a bank teller and an activist in the feminist movement. But in real life, humans rely on context in understanding the meaning of questions. The word probable for instance has multiple meanings. In the mathematical sense it is synonymous with frequency. Interpreted in this way, it yields the 'correct' answer noted earlier. But probable is also used in the nonmathematical sense as referring to what is plausible or what there is evidence to support. This comes closer to the way probability is used in assessing guilt in a court of law. Furthermore, most human interactions involve some judgements about the relevance of information in a decision context. So, an individual considering Linda might make the contextual judgement that because the information about Linda is provided by the questioner it is likely to be relevant, and it certainly provides evidence that would be supportive of the judgement that she was an activist. From this perspective, the artificiality of the context of the experiment is what makes the alternative answer so clearly wrong.

8.10 Some advantages of heuristics

Given that so many real-life human problems require a response in a limited timeframe and must be based on very incomplete information, it is not surprising that often an appropriate heuristic is more useful than an attempted logical and probabilistic calculation. It can also often be more likely to be correct. Consider the following question: Which city has more inhabitants, Detroit or Milwaukee? When this question was posed to a group of undergraduates at the University of Chicago, about 60% chose the correct answer,

Detroit. When the same question was posed to a group of German students, nearly all of them got the correct answer. How can this be? Surely German students would know less about the population of American cities than American students. Upon examining the basis for their choices however the answer became clear. The American students tried to draw on their knowledge of the two cities and made their choices accordingly. The German students in general had heard of Detroit but had not heard of Milwaukee. Hence, they guessed that the city that they recognized was more likely to be larger than the one they did not. In short, they used a 'recognition heuristic'[21]: if one of two objects is recognized and the other is not, then the recognized object has a higher value with respect to the criterion in question. Obviously, the recognition heuristic is not infallible. However, it is successful a remarkably high percentage of the time, because in real-life situations recognition is often strongly correlated with the criterion in question. This highlights the value of the heuristic being frugal, that is requiring very little information for it to be applied successfully. The old medical adage 'If you hear hoofbeats, first think of horses, not zebras' is an example of the recognition heuristic.

Let us consider another example of a heuristic discussed by Gigerenzer, the 1/N rule.[22] An important problem in economics is the issue of optimal asset allocation: how to best invest one's assets among a set of potential options, such as stock funds. Numerous economists have developed complex models for how to make this decision based on considering the past performance of the various funds in question over a given period of years. A study compared the application of 12 such complex models over seven different allocation problems. Interestingly none of the models could beat a simple heuristic of dividing one's assets equally among all the investment vehicles being considered, that is, put 1/N of your assets in each vehicle.

How can this be? It has to do with the issue of robustness. Complicated models did better than the simple heuristic in data fitting, that is adjusting their parameters to the data over a past number of years. However, they do worse in predicting the future behaviour of the investment vehicles. In looking at this problem it becomes clear that the larger the uncertainty and the number of assets being considered and the shorter the timeframe of past performance being considered, the greater the advantage for the heuristic over complex models. It was estimated in this study that if one were looking at 50 potential investment vehicles one would need an accumulation of 500 years of performance data before any of the complex models would outperform this simple heuristic.

Analogously in psychiatry, for a depressed patient it is not feasible to sequentially try every antidepressant, alone and with an augmenting agent, to find the best pharmacological treatment for that patient. Instead, the treating psychiatrist applies a set of heuristics that heighten the likelihood of finding an adequately effective and tolerated medication(s) in a reasonable time frame. Some are part of the general clinical wisdom of the field and some uniquely derived from that psychiatrist's own experience. Some of these may relate to expected tolerability; e.g. selective serotonin reuptake inhibitors (SSRIs) before tricyclic antidepressants. Others may relate to certain patient characteristics that the psychiatrist feels may suggest differential response; e.g. SSRIs for an anxiously depressed patient and bupropion for an anergic one.

8.11 Optimizing and satisficing

The research programme of the science of heuristics has studied a wide array of other specific heuristics with equally intriguing results. For our purposes, though, I want to use this distinction between judgements based upon logical and probabilistic calculations and heuristic strategies to consider general approaches to human decision-making and deliberation. The paradigmatic case of a human decision involves selecting one from a small number of possible actions to address a complex life situation. Such situations involve multiple factors that could influence the outcome of the decision and hence count as reasons for or against some or all the feasible options. The factors vary in importance and their relative degree of independence from one another.

Two basic families of strategies can be applied in such a situation, and are generally termed 'optimizing' versus 'satisficing' approaches.[23] The latter neologism was coined by Herbert Simon, an early pioneer in this field of research.[24] Optimizing strategies work on the assumption that there is in theory a best or optimal answer. If one had all the necessary information in sufficient detail and enough time, one could calculate the best answer leading to the best decision. Given that in real-life situations we seldom have all the information in sufficient detail, optimizing strategies are designed to deploy some sort of calculus to give us the highest probability of that correct answer. Many models have been proposed to achieve this. What they typically have in common is an attempt to identify all the relevant factors that could count as reasons, to have some strategy for quantifying their weight or importance for the decision at hand, to have a mechanism for accounting for their relative dependence or independence from one another, and some calculus for combining all the factors to generate the optimal decision.

Even in fields where data tends to lend itself to quantification, such as economics, optimizing strategies are daunting. The formulas are complex and require making many assumptions about the nature of the data and their relationship to one another. Values for relevant data are seldom precise and must be approximated, sometimes quite crudely. Estimations as to the quality of the data must be made. The outcomes are highly sensitive to small variations in the data, making it difficult to generalize from one situation to another. When optimizing strategies are applied to other day-to-day decisions where information is less intuitively quantifiable, all these problems are magnified immensely. Furthermore, as the number of variables increases these kinds of calculations rapidly become intractable. Such optimizing approaches have much in common with statistical regression models.

In short, for many real-life human deliberations, optimization approaches are simply impractical and intractable. Furthermore, they are difficult to apply in situations where there is more than one competing goal. Beyond this, optimization does not imply an optimal outcome. As Gigerenzer notes:

> . . . the term *optimization* refers to a mathematical process—computing the maximum or minimum of a function—which does *not* guarantee optimal outcomes in the real world. The reason is that one has to make assumptions about the world in order to be able to optimize. These assumptions are typically selected by mathematical convenience, based on simplifications, and rarely grounded in psychological reality. If they are wrong, one has to build the optimization castle on sand and optimization will not necessarily lead to optimal results.[25]

8.12 Two heuristics for decision-making

How in practice, then, do human beings make day-to-day decisions both trivial and important? Studies of actual human decision-making shows that several decision-making heuristics are commonly employed. None of these heuristics can claim to produce the best or optimal decision. Rather what they seek to do is to generate a course of action that is adequate to solve the problem at hand within the time frame available and without an excessive use of resources to gather more and more information. It is this idea of generating a solution that is adequate if not optimal that is meant by calling it a 'satisficing' strategy. This does not mean that the solution will not be the best option. In many cases it will be; but a satisficing solution is not derived by a strategy that is designed to assure the optimal solution.

There are two common classes of heuristics that are frequently employed in these situations when one is looking at multiple reasons for and against a decision.[26] The first is often referred to as a 'take the best strategy'. In this approach one starts with a list of possible reasons for or against the options in question. For simplicity we will assume a case with two options, A and B. The reasons are listed in order of importance from the most important to the least. A binary decision as to whether the reason argues for or against the options is made. One starts with the most important reason. If it argues for A and against B, A is chosen and the deliberation is over. If it argues for both options or against both options, the deliberator goes to the second most important reason and asks the same question. The process continues until some reason discriminates between the two options, at which point that option is chosen and the deliberation ends.

The second common heuristic for this sort of decision-making is called 'tallying'. In this case a finite number of relevant potential reasons are postulated. A somewhat arbitrary and impressionistic decision is made as to how many of these reasons need to be present before one is willing to opt for either A or B. Then one examines each potential reason in turn, not necessarily in any predetermined order. Once either option accumulates enough positive reasons, the decision is made and the process ends.

Both heuristics can be applied in the absence of precisely quantified data and in the face of considerable unknowns. They can be implemented in a short period of time. They are much less disrupted by missing information and do not require forcing variables into quantitative form when they do not naturally fit; i.e. pseudo-quantification. In short, they are frugal and robust. What may be more surprising is that researchers of the heuristic school have tested the outcomes of these strategies against quantified optimizing strategies, such as multiple regression, in a wide range of human problem-solving situations. In many of these situations these heuristics outperformed multiple regression.[27] This likely reflects the fact that in real-world situations so many assumptions must be made in multiple regression models to account for the absence of precise data that the models become highly inaccurate. Furthermore, the requirement for precise data often means that these kinds of models do a poor job predicting a new situation, as even small variability in data points changes the outcome dramatically. In short, they are not robust.

As I reflected on these issues, I realized a situation in which I have long tended to employ a 'take the best' strategy without consciously realizing that I was employing that heuristic. In a patient with major depression who is a candidate for an antidepressant, I always ask in taking the history not only whether other first-degree relatives have suffered from depression, but whether they showed a particularly good response to any particular antidepressant. If a relative did so, I almost invariably use that antidepressant for my initial treatment trial with that patient, barring some compelling reason to avoid that medication in this individual. I realized that I assume, rightly or wrongly, that there is far more predictive value from this one bit of information than from trying to weigh a large number of considerations given that there is little evidence for strong differential efficacy between antidepressants.

One of the projects of the heuristic research programme is to identify those environmental features of a particular situation calling for a decision that make a particular heuristic more or less appropriate. Thus, it is reasonable to ask when it makes sense to use a 'take the best' strategy versus a 'tallying' strategy. Studies[28] have shown that when there are large differences in the relative importance of different potential reasons, a 'take the best' strategy is preferable. This makes intuitive sense in that the most important reason is likely to be considerably more important than the second or third most important reason. Hence stopping at the first discriminating reason is likely to capture the bulk of the variance. Conversely, when reasons are perceived to be of relatively similar magnitudes of importance, a 'tallying' strategy appears best. Human decision-makers are often quite good at deciding which situation applies without having precise quantitative weights for the various reasons. Intuitive judgements usually serve sufficiently well. These decisions as to which heuristic to apply are typically made unconsciously and quickly. People of course vary in their competence in choosing the correct heuristic. Studies have shown that people with higher IQs are better at detecting the relevant environmental features and consequently at knowing which heuristic to use.[29] It is also clear from experimental work that competence at heuristic selection over a range of situations can be learned by teaching and experience, so long as feedback is provided.[30]

8.13 Heuristics and expertise

What happens to the selection and use of heuristics in decision-making as one develops some expertise in a field? The researcher who has spent his career exploring this question is Gary Klein. His central findings and theoretical understanding about expertise is summarized and illustrated in his excellent book, *Sources of Power: How People Make Decisions*.[31] We will look closely at Klein's work, both because it is informative about how expertise develops and operates and because his formulation is so strikingly parallel to POP modelling. Klein was struck by the fact that in many fields experts are called upon to make important and critical decisions extremely rapidly and often with limited information. The goal of his research project has been to understand the thought processes of such experts, both to explain the remarkably high proportion of the time their decisions are correct as well as to develop techniques for teaching expertise to novices entering the field. His earliest studies focused on critical decision-making in situations with high-stakes, extremely constrained time frames, and limited and ambiguous information. Among his early subjects were experienced firefighters, fighter pilots, and military squad leaders. His study approach was naturalistic and based upon detailed semi-structured interviews with the experts in question to elicit an in-depth understanding of their thinking processes. Rather than confronting them with artificial laboratory situations, the interviews focused on explicating their decision-making process in actual cases the experts had confronted.

One characteristic finding was that the experts often could not initially state the reasons for their decision, seeing it as an intuitive or gut decision about which nothing else could be said. As the interviewer encouraged the experts to go over the details of the situation more explicitly, they were usually able to identify features of the situation that informed their decision. What became clear was that the experts had developed through their experience a set of patterns or paradigms of typical situations that they had encountered in the past. Confronted with a new situation, the expert had a rapid intuition of one of two sorts. Either the situation was quickly recognized as fitting a known pattern, in which case the course of action would follow relatively directly. Alternatively, certain features of the new situation violated the relevant paradigm suggesting the presence of some anomaly or novelty in the current situation. This could in turn require an altered course of action.

8.14 The recognition-primed decision model

Klein came to view this initial step in exercising expert decision-making as central. Hence, he termed it the 'recognition-primed decision model' (RPD). It starts with an active intuition. The intuition is not some mysterious process, but 'depends on the use of experience to recognize key patterns that indicate the dynamics of the situation. Because patterns can be subtle, people often cannot describe what they noticed, or how they judged the situation is typical or atypical.'[32] A classic example in the psychiatric literature is the *praecox gefühl*, a powerful but vaguely impressionistic sense the experienced clinician has that suggests one is in the presence of someone suffering from a schizophrenic disorder.[33] When pressed, the clinician can identify certain attributes of the patient and the mode of interaction with the clinician that form the basis of this intuition.

Klein goes on to argue that intuition comes from accumulated experience, but only if that experience has certain attributes. The individual must be exposed to many experiences that are both typical and atypical to develop an appreciation for common patterns as well as indicators of anomaly. Furthermore, the individual must get feedback as to the outcome of the situation in a short enough timeframe so that one learns what patterns of response typically work or fail to work.

I had an early experience that illustrates this process. As a college undergraduate, I worked as an aide at Chestnut Lodge, a renowned if controversial private psychiatric hospital in the state of Maryland in the US, that used psychoanalytic approaches for treating seriously ill patients with minimal use of medications. Behavioural control of aggression was a major concern. One of the psychiatrists there told me that the staff had noted that whenever a particular aide was assigned to a unit, the rate of aggression was dramatically and consistently lower during his shift. He had worked at the hospital for many years, had only a high school education and no prior formal psychiatric training. When told about this pattern, the aide was unable to identify anything specific in his approach that might explain his success in reducing aggression. The psychiatrist then reviewed with him in detail several specific instances when he successfully de-escalated a potentially aggressive episode with a patient. The psychiatrist was even able to capture a few of these encounters on video and had the aide watch them, inviting him to comment on his own decisions. Eventually, they were able to identify certain cues that alerted the aide to impending aggression early in the process, as well as characterizing some successful strategies for de-escalating these situations. The aide had been totally unaware of having any specific cues or strategies, and was fascinated seeing his own expertise becoming consciously articulated. From all this, a training module was developed for use with the entire clinical staff of the hospital.

While having a wide range of experience is critical for developing expertise, Klein argues that there are instructional strategies that can facilitate and speed up this process.

> The part of intuition that involves pattern matching and recognition of familiar and typical cases can be trained. If you want people to size up situations quickly and accurately, you need to expand their experience base. One way is to arrange for person to receive more difficult cases . . . Another approach is to develop a training program, perhaps with exercises and realistic scenarios, so the person has a chance to size up numerous situations very quickly. A good simulation can sometimes

provide more training value than direct experience. A good simulation lets you stop the action, backup to see what went on, and cram many trials together so a person can develop a sense of typicality. Another training strategy is to compile stories of difficult cases and make these the training materials . . . Perceptual clues can be defined. [Looking at results from a number of training programmes] . . . emphasis on pattern matching seemed more useful than lessons on formal analysis of alternate options.[34]

8.15 The role of simulation in choosing an action

The exercising of intuition to determine whether an instance is an example of a known pattern or is anomalous in some way is just the first step for the expert. This intuition must lead to a course of action. How is that action decided? As he began his initial studies, Klein expected that experts would use some rapid and efficient heuristic where a small number of options would be considered and quickly assessed to allow a rapid choice. To his surprise he found that experts typically employed an even more rapid decision-making process that he came to call 'mental simulation'. This he defines as 'the ability to imagine people and objects consciously and to transform those people and objects through several transitions, finally picturing them in a different way than at the start'.[35] This is done by positing a manageable number of states and transitions between the states, all of which constitutes what Klein calls an 'action sequence'.

Such simulations can be used to both explain the past sequence of events that lead to the current situation and to project into the future; that is, how to move from the current situation to another more desirable one. By mentally going through the action sequence, the expert can imaginatively play out how a particular decision option is likely to go; i.e. will it lead to the desired result or likely hit some snag requiring adjustment. Klein found that experts especially under conditions of high time pressure, did not consider even a small number of options by weighing the pros and cons. Rather they imaginatively jumped to a potential course of action and then played out the mental simulation to see if it brought about the desired goal. If it did, they attempted to implement it. If the mental simulation did not bring about the desired goal, they then generated a second option that in turn was subjected to mental simulation. He found that in situations of high time pressure,

. . . the mental simulations were not very elaborate. Each seem to rely on just a few factors—rarely more than three . . . Perhaps the limits of our working memory had to be taken into account. Also, there was another regularity: the mental simulations

> seem to play out for around six different transition states, usually not much more than that. Perhaps this was also a result of limited working memory . . . Of course, there are ways of avoiding the constraints. If we have a lot of familiarity in the area, we can chunk several transitions into one unit. In addition, we can save memory space by treating a sequence of steps as one unit rather than representing all the steps. We can use our expertise to find the right level of abstraction . . . Another strategy for overcoming memory limits is to write things down and draw diagrams to keep track of the transitions . . . Considering all these factors, the job of building a mental simulation no longer seems easy. The person assembling a mental simulation needs to have a lot of familiarity with the task and needs to be able to think the right level of abstraction. If the simulation is too detailed, it can chew up memory space. If it is too abstract, it does not provide much help . . . without a sufficient amount of expertise and background knowledge, it may be difficult or impossible to build the mental simulation.[36]

Of course, in situations where the expert has more time, the mental simulation models can become somewhat more elaborate. However, after a certain level of complexity, they become increasingly unmanageable, even for the expert. Hence achieving the right level of abstraction remains critical.

Klein fully understands that mental simulations are not infallible. In fact, he looks at several of the ways they can mislead. After we develop a mental simulation to reflect the situation, there is a tendency to discard disconfirming evidence. 'Once we have built a mental simulation, we tend to fall in love with it. Whether we use it as an explanation or for prediction, once it is completed, we may give it more credibility than it deserves, especially if we are not highly experienced in the area and do not have a good sense of typicality'.[37]

Klein points out that several strategies have been developed to help the individual look more critically at their own simulations. While he notes that mild discrepancies may best be explained away since the evidence itself is not always trustworthy, as more and more discrepancies appear and the simulation must be adjusted in ever more complicated ways, this very complexity should raise a red flag as to whether another simulation would be more appropriate. Another strategy that can be used to make people more aware of the possible inadequacies of the simulation is what Klein calls a 'premortem' exercise. Here one is asked to imagine that the simulation has been implemented and, at some point, it has failed to bring about the desired outcome. Then one can ask what the most likely sources of the failure could be. Hence one is asked in advance to look for the potential weak links in the simulation.

Klein has demonstrated in his work that as an individual becomes more of an expert in any field their decision-making pattern shifts from various strategies of weighing multiple options, including attempts at optimization, to this RPD model characterized by intuitive pattern recognition with the judgement of typicality or atypicality, followed by mental simulation. If this is how expertise works, then it follows that the training of future experts should look to develop RPD as quickly and effectively as possible.

8.16 POP model construction as simulation

Beyond some differences in terminology, the process of mental simulation is essentially synonymous with the process of model construction that we elaborated in considering POP models in psychiatric reasoning. In both cases experts begin by looking for recognizable patterns in the problematic situations they are confronting. If the pattern is a highly familiar one from other situations in one's experience, the psychiatric expert is likely to formulate an understanding of the situation in terms of a mental structure (POP model) characterized by a number of elements and how they interrelate (interacting propensities in POP language), and to do this so rapidly that it appears to be a spontaneous intuition. Also, a course of action (proposed interventions) is likely to present itself equally rapidly. The case of Ms K is sufficiently straightforward and prototypical to approximate this situation, as is the initial model for Mr Z. If there are more anomalous elements present or the case is less prototypical, the expert is likely to be more aware of the cognitive process at work in generating the model. This is true for the revised model for Mr Z, and the cases of Mr D and Ms M, where considerable conscious deliberation was required in constructing the models.

The next step by the expert is to use the model and the proposed course of action or interventions to mentally simulate how things are likely to change when the interventions are made. The speculation as to differential responses to antidepressants for Mr Z is an instance of this. If all proceeds well with the mental simulation, the interventions or action sequence will be initiated. If a problem appears, the model will be revised or replaced with a better one. If in the real-life execution a roadblock appears or the desired outcome is not forthcoming, the model will again be revised and the process is repeated. The initial failure of my psychotherapy with Ms M is an example here. POP modelling then is an instance of how experts make decisions in a wide range of disciplines where ambiguity, incomplete information, and the need to make decisions under time pressure are present.

Our next step then is to take what we have learned about human decision-making and expertise in this chapter and look more directly at how it applies in psychiatric practice and training. That will be the focus of the next chapter.

Endnotes

1. Hacking 2006, Weisberg 2014
2. Weisberg 2014
3. Press 1999
4. Gigerenzer 2008; Kahneman 2011; Klein 1998
5. Jaeger 1957; Ierodiakonou 2014
6. Aristotle 1999
7. von Neumann and Morganstern 1944
8. Kahneman 2011
9. Kahneman 2011, pp. 26–28
10. Kahneman 2011, pp. 44–46
11. Kahneman 2011, pp. 41–44
12. Kahneman 2011, pp. 156–165
13. Kahneman 2011, pp. 421–422
14. Kahneman 2011, pp. 109–188
15. Gigerenzer 2008
16. Gigerenzer 2008, p. 8
17. Gigerenzer 2008, pp. 21–22
18. Gigerenzer 2008, pp. 130–131
19. Hoffrage *et al.* 2000
20. Gigerenzer 2008, pp. 69–73
21. Goldstein and Gigerenzer 1999
22. Gigerenzer 2008, pp. 9–10
23. Gigerenzer 2008, p. 114
24. Simon 1955
25. Gigerenzer 2008, p. 84
26. Gigerenzer 2008, pp. 32–33
27. Gigerenzer 2008, p. 41
28. Gigerenzer 2008, pp. 36–38
29. Gigerenzer 2008, pp. 38–39
30. Rieskamp and Otto 2006
31. Klein 1998
32. Klein 1998, p. 33
33. Goze *et al.* 2019
34. Klein 1998, pp. 45–47
35. Klein 1998, p. 49
36. Klein 1998, pp. 56–57
37. Klein 1998, p. 71

Chapter 9
The role of model construction in developing expertise in psychiatry

9.1 Summary of expert decision-making

In the previous chapter we conducted a rapid survey of attempts to understand human reasoning in general, culminating in a consideration of current theories about human decision-making and expertise. In making real-world decisions, there exists no infallible way of reaching the guaranteed best decision. This makes sense given that human decisions occur in a context of less than complete and fully accurate information. Furthermore, the complexity of factors involved in most real-world decisions make precise calculation of the correct answer intractable. Research has shown that humans do not act as embodied calculating machines rigidly applying algorithms or formal rules of logic and probability. Rather we apply certain cognitive strategies to achieve an acceptable answer in a reasonable timeframe. Possible strategies fall into two categories, optimizing and satisficing. Optimizing strategies attempt to come as close as possible to the calculating machine model by attempting to consider all the relevant factors, appropriately weighted, with the goal of achieving the best possible answer. Due to the practical limitations noted in the last chapter, this process is often cumbersome and indeed intractable. It is also easily skewed by missing and variable information, so that the results are often in fact unsatisfactory.

More recent studies have shown that most human decision-making relies on much simpler heuristics that do not seek to guarantee the theoretically best decision, but rather one that is sufficiently good to work for the situation at hand. The work of Klein and others has argued that novices in any field tend to use strategies that resemble the more cumbersome optimizing approaches but that as one obtains greater expertise, simpler heuristics are employed and tend to culminate in an even more streamlined process that he terms recognition-primed decision-making (RPD). This is clearly a satisficing strategy. The goal is to find a course of action that works for the problem at hand. It leaves open the question of whether better strategies

How Psychiatrists Make Decisions. Douglas W. Heinrichs, Oxford University Press. © Oxford University Press (2025).
DOI: 10.1093/9780198960720.003.0009

might exist. The strategy has two primary components: pattern recognition and mental simulation. The first involves the ability to recognize prototypical patterns, a capacity that grows with one's experience in the field, and to recognize anomalies that do not fit a pattern and that require strategic adjustments. When a prototypical pattern is recognized, it usually occurs rapidly and hence often tacitly incorporates some notion of the causal factors involved in the action that the expert will take. Where anomalies are present some reflection may be involved in postulating the relevant causal factors and their interactions leading to an action plan. Mental simulation, in turn, takes the pattern of causal factors and the resulting action plan and plays the simulation mentally, and if it appears to work the expert will proceed to implement it. If the simulation suggests a problem or roadblock, the simulation is modified, or if that is not possible, rejected, and a new action plan is developed and another simulation run. Once the expert begins to conduct the action plan, the emergence of unexpected problems or obstacles can again lead to postulating a modified or new pattern of causal factors and a new action plan that in turn is subject to simulation.

9.2 Applying research on expertise to psychiatric education

Psychiatrists, typically have more time than the firefighter or the military commander to weigh and make decisions. Nonetheless, as expertise develops, RPD would be expected to be relied on more heavily. Experienced physicians, in evaluating a new patient or problem, form diagnostic impressions, and consider likely treatments extremely early in the evaluation process, often within minutes. Much of the rest of the interview and assessment involves collecting information that either supports or challenges that initial impression. If this is indeed how people function with growing expertise in a field, strategies to improve RPD should improve an expert's performance as well as facilitate the growth from novice into expert. Klein has argued that such strategies exist and can be employed in the development of expertise in any field. In this chapter I want to take up this challenge as it applies to developing and enhancing expertise in the practice of psychiatry. This will apply equally to efforts by psychiatrists who already has some degree of expertise who wishes to improve their performance, as well as strategies for the training of psychiatrists early in their career to enhance their transition into expert practitioners.

I suggest that the explicit use of pattern of propensities (POP) models can play a valuable role here. My original formulation of POP models in psychiatry came from an attempt to understand how psychiatrists reason in making decisions about the treatment of individual patients by reflecting on my own work with patients. It is striking how this POP model in fact overlaps with the description of RPD as developed by Klein. In section 8.16, I argued that, despite some minor differences in the descriptive language, they are in fact synonymous. The way a psychiatrist uses a POP model is an instance of an expert engaging in pattern recognition, hypothesizing a pattern of propensities and their interactions, proposing actions sequences, and then subjecting those to mental simulation. The propensities and their hypothesized patterns of mutual influence equate to the causal factors and transition states of the mental simulation. The resulting action sequence corresponds to the hypothesized therapeutic interventions of the POP model. Using POP models to enhance the training of psychiatrists and to help novices become experts constitutes the pedagogical function of this approach.

9.3 Pattern recognition and families of models

It may at first glance seem paradoxical to talk about pattern recognition by the psychiatrist when we have emphasized that models are unique to the individual patient. The resolution of this paradox will be easier to grasp after the discussion in section 10.2.3.2 of the role of models in the contemporary philosophical understanding of how science in general is practised and the idea of families of models. Briefly, at the most basic level a concrete model is generated and used by a scientist to solve a specific problem by representing a specific piece of the real world. As experience accumulates in dealing with many problems, certain common elements in some of these unique models become apparent. This makes it possible to develop more abstract models that characterize similar groups of unique, concrete models. When scientists confront a new problem that resembles one of these groups, they can use the abstract models to suggest elements that might be useful in constructing a new concrete model to deal with the new real-world problem. These more abstract models also serve as pedagogical tools that can be taught to novices to help them construct more concrete models of problem situations even before they have developed their own body of experience. As experience continues to grow, the abstract models themselves will appear to fall into certain groups with similarities. Each of these groups in turn can be characterized by an even more abstract model. The philosopher of science Ronald Giere[1] characterizes this as families of families of models.

A psychiatric illustration may be clarifying here. How did I go about constructing the initial model for Mr Z? He presented with a range of depressive symptoms as well as anger. In constructing a POP model unique to him and his situation, I considered a range of more generalized or abstract understandings thought to characterize groups of patients who share depressive and anger symptoms. These more abstract models are derived from a range of sources, including naturalistic and controlled studies of depressed individuals, basic science research on depression, psychological theorizing from psychodynamic, cognitive-behavioural, or other frameworks, as well as less formalized collective clinical wisdom shared in one's training. In addition, I was able to consider the range of my own experiences in dealing with other patients with similar symptom presentations. In cases of depression, for instance, one can consider the presence of early loss and trauma, genetic factors, negative cognitive distortions, early interactive patterns with significant others leading to shame and low self-esteem, and putative differences in neurotransmitter action characteristic of depression. From these elements found in more abstract models of the 'depressed person', the clinician can choose those factors judged to be applicable to the specific, concrete patient for whom one seeks to generate a POP model.

It may be possible to generate even more abstract models. For instance, some of the factors common to many individuals with depression may be shared with individuals with other kinds of disorders, such as anxiety disorders, allowing the creation of a still more abstract model of the 'person with an internalizing disorder' that would contrast with an abstract model of the 'person with an externalizing disorder' where a different set of propensities and their interactions apply. At the most abstract level are the various theoretical frameworks in terms of which all the hierarchy of models can be cast. In the psychodynamic framework, for instance, the clinician employs very general processes such as defence mechanisms, the unconscious, transference and countertransference, and stages of psychosexual development to formulate specific models. From a cognitive-behavioural perspective, pathways by which negative cognitions affect feelings and behaviour are central components in formulating specific models. From a biological perspective, principles of neurotransmitter and other neuromolecular mechanisms serve as organizing principles for model construction.

In fields like psychiatry, even at this most abstract level there are competing models. This is generally true in the social sciences as well. In more matured natural sciences, like physics and chemistry, a single model tends to dominate at the more abstract levels. Yet even in physics, the most generalized and abstract models, such as classic mechanics and quantum theory, have yet to be successfully integrated.

It is the most concrete level of models that map on unique real-world individuals or situations. The more abstract levels of models are not actually models of entities that exist in the world, but are theoretical tools that serve the purpose of providing some guiding notions that can be useful in constructing new concrete models to cover new unique individuals and situations. There is no 'the depressed person' or 'the person with internalizing disorder'. These terms identify theoretical models, whose only value lie in helping us construct specific models of the individual persons we are attempting to understand and help.

Like RPD, POP modelling is clearly a satisficing strategy, not an optimizing one. It leaves open the question of whether a better strategy exists. If the therapeutic interventions resulting from the model lead to an adequate resolution of the patient's difficulties, it is successful. If not, it requires revision or rejection. The contrast between optimizing and satisficing strategies provides another perspective for appreciating a fundamental difference between POP modelling and approaches advocated by evidenced-based medicine (EBM). EBM, in its most ambitious forms, is clearly an attempt to apply an optimizing strategy. The goal is to try to determine the very best intervention for this patient based on all available information in the literature, with a special emphasis on controlled trials. It believes that even though our current knowledge is incomplete, at any given time there exists a best answer to the question of how to treat Mr Jones. As has been argued in section 2.7, the effort to do this has led to an extremely cumbersome and unwieldy process that is forced to make many questionable assumptions and impose arbitrary pseudo-qualifications in its attempt to move from the results of studies of groups of patients to the application to a given individual. As such EBM demonstrates the characteristic problems with applying optimizing strategies in the real world.

9.4 Core elements of psychiatric training

Given all this, how should we think about the necessary elements in the training of a psychiatrist that would enhance the development of expertise? I would suggest that they can be grouped into four categories:

1. To have a pool of potential elements from which to draw an understanding of a patient and know what effective strategies are potentially available for any given patient, the novice needs a knowledge

base in a range of informing disciplines including psychopathology, therapeutics, neuroscience, psychology, and sociology.

2. The novice must develop the right sort of experience base, and reflections on that experience through prompt feedback, to acquire the intuitive grasp of prototypical as well as anomalous patterns. This presumes the skills to effectively collect information about the patient's unique experiences. This includes obtaining a good psychiatric history, conducting a thorough mental status examination, establishing a working relationship with the patient, and obtaining adequate follow-up as to the outcome of interventions.
3. The psychiatrist in training needs to develop the ability to generate appropriate POP models to guide deliberation, and then to execute effective mental simulations with those models leading to an appropriate course of therapeutic action. This includes learning how to recognize when the simulation will likely work or, once implemented, to recognize that it is not working, and take appropriate corrective action. Part of this process is learning to minimize potential blind spots and biases.
4. The psychiatrist in training must develop skills in implementing the various therapeutic interventions. This specialized skill set is most obvious in the case of psychotherapies, but many skills apply to the effective application of pharmacotherapy as well. The unique contribution that comes from explicit attention to POP models and the process of RPD apply primarily to categories 2 and 3. Hence, they will be the primary focus of this chapter. But first, several brief observations about categories 1 and 4 are in order.

9.5 Acquiring basic knowledge and the skills to implement interventions

Category 1 relates to acquiring a solid knowledge base from other informing disciplines and sciences. Meeting this need was the original goal of EBM. EBM started from the realization that traditional ways of acquiring knowledge in the rapidly expanding informing disciplines for psychiatry were woefully inadequate. Textbooks and even journals are so quickly out of date. More new information is generated than can possibly be read let alone critically assessed by the busy clinician. The initial mission of EBM was to train clinicians in how to locate the most current information efficiently as it related to a specific clinical problem and how to critically assess the quality of that information. This is a laudable and important goal. None of my critiques of EBM apply to this aspect of it. The problem for EBM arises when it was expanded from addressing the challenges of this category to attempting to provide a way of dealing with category 3. The elaborate edifice of later EBM work is an attempt to transform it into a comprehensive strategy for making optimal clinical decisions about an individual patient. It is only this overreach that represents a conspicuous failure. The original mission of EBM is more than challenging enough, and should remain the focus of its efforts.

Component 4 involves acquiring the skills to implement appropriate treatment interventions once they have been decided upon. There are many questions and controversies in this arena that lie outside the focus of this book: Given the wide range of treatment interventions currently available, both biologic and psychological, can the psychiatrist be expected to develop expertise in delivering all of them? If not, which ones? In an era where split treatment by multiple providers has become the rule rather than the exception, how should treatments be best delegated and coordinated and who should make those decisions? Who has the responsibility to look at the big picture and assess the effectiveness of the integrated treatments? How does one best collaborate with other providers? However one chooses to answer these questions, it seems indisputable that it is incumbent on each provider, including the psychiatrist, to develop the skills to establish a supportive and empathetic working relationship with each patient that considers the unique set of emotional needs and characterological idiosyncrasies. The psychiatrist should also have enough understanding of the full range of treatment modalities to know when they are appropriate for each patient, even if one does not choose to implement the interventions oneself.

Let us now turn to the other two categories of elements important in developing expertise in psychiatry: 2) acquiring sufficient experience of the right sort to facilitate recognizing prototypical and anomalous patterns and 3) developing POP models and then subjecting them to effective mental simulations before initiating interventions, as well as assessing in an ongoing way the effectiveness of actual interventions. I suggest that these two categories have not been optimally addressed in traditional psychiatric training.

9.6 Acquiring experience for the development of psychiatric expertise

The novice psychiatrist must develop, through the right sort of experiences and reflections on those experiences, the capacity to accurately recognize both prototypical and anomalous patterns that directly lead to a hypothesized course of therapeutic action. Psychiatric residents are certainly exposed to a wide range of patients during their training and are required to think about what potential therapeutic interventions flow from their grasp of the patient. However, as the work of Klein suggests, conscious attention to this process can lead to strategies that markedly improve the quality of this activity. I believe that several typical aspects of traditional psychiatric training constitute impediments here.

In many settings the assumption is made that the sheer accumulation of experience will lead to this skill set. This experience may take the form of the psychiatrist in training wrestling with her own cases or observing the conclusions reached by more experienced clinicians, such as senior residents and supervisors. All too often however the latter takes the form of the more senior clinician stating an understanding of a given patient with little attempt to articulate the basis for that judgement or considering the ways it may be wrong. In part this reflects the reality noted by Klein in his studies of experts that, without the help of an interviewer or a conscious effort at guided self-reflection, the expert often has little or no conscious understanding of one's own cognitive processes. As a result, these intuitions can take on a mysterious and magical quality, as if with enough experience some occult skill is acquired. This is reflected in the frequent use in the unfortunate phrase 'the art of medicine'. This sort of belief, often unconsciously held, is frequently shared by both the novice and the supervisor. I remember in my own training being supervised by a very senior and nationally renowned psychiatrist who, in answer to my questioning the basis for a judgement he had made about one of my patients, referred only half-jokingly to his 'unimpeachable clinical judgement'.

A second factor in current psychiatry that frequently works against the acquiring of meaningful pattern recognition with an individual patient is a tendency to assume that establishing the diagnosis equates with recognizing the relevant pattern. As discussed at greater length in section 7.15, in developing the POP model a diagnosis is at most one piece of the pattern articulated in a model. On the one hand many features of the pattern instanced by an individual patient involve elements unrelated to the diagnostic category. On the other hand, our current diagnostic categories are symptom-based and attempt to be atheoretical. As a result, even within those factors included in the diagnostic category, multiple potential patterns exist that would all place a patient within the same diagnostic group. Finally, preoccupation with diagnostic considerations can lead to neglect of elements that might be unique to a given patient or a small subset of patients within a diagnostic group.

This use of diagnosis as a substitute for more comprehensive pattern recognition, combined with the uncritical application of EBM-based treatment decisions derived from groups of patients can lead to a remarkable blindness about an individual patient's pattern, especially where an anomaly exists. One way of understanding what happens here is that the process of pattern recognition characteristic of an expert is cut short. It is important to remember that the recognition of a pattern implies some understanding of the individual instance in terms of a model of relevant propensities and their interaction and implies the notion of a course of action. It is this course of action that, given the model, can be subjected to the process of mental simulation to test out its likely effectiveness. If the potential courses of action are simply responses to individuals having a particular diagnosis, they are cut off from what is unique about specific patients, and the complex situation in which they are embedded and through which they are attempting to navigate. This is especially true where anomalies exist.

9.7 An example of diagnosis-induced blindness

Just how extreme the resulting blindness to individual circumstances can become was brought home to me several years ago when I was asked by a family to do an independent psychopharmacological consultation on their mother who was hospitalized in a well-regarded private psychiatric facility.

Ms W was a 66-year-old married woman with three grown children. She had a history suggesting a quite severe generalized anxiety disorder that was probably lifelong, but certainly existed from at least early adulthood. She functioned well as a wife and mother but suffered persistent distress from her anxiety. In her mid-30s her primary care physician prescribed diazepam at a dose of 2 mg three times a day. The results were profound and life altering for Ms W. Her anxiety dropped dramatically to quite normal levels. This was noticeable both to her and her family. She continued the same dose of diazepam for three decades with continued excellent response. She tolerated this modest dose without any evidence of side effects. No dosage escalation occurred over the years nor was there any evidence of drug-seeking behaviour, abuse, or diversion. Ms W had no personal or family history of substance abuse.

Her current problems began about nine months before I was asked to see her. Her long-term primary care physician retired and sold his practice to a young internist recently out of training. At her first visit with her new doctor Ms W was told that it was inappropriate and dangerous for her to have been on diazepam for so many years. The doctor told her that she was addicted and that she had undoubtedly developed tolerance so that it was no longer doing anything for her. He told her that it was necessary to begin a taper off her diazepam, especially as her age was a contraindication to its use. He conducted the taper extremely slowly and conscientiously. However, by the time she was off diazepam she was struggling with constant overwhelming anxiety. She felt a constant state of inner agitation reflected in pacing and some tremor. She complained of tightness in her chest and shortness of breath. Her sleep was disrupted. She struggled to focus on her day-to-day tasks. When she described this to her physician, he told her that the current treatment recommended for chronic generalized anxiety was a selective serotonin reuptake inhibitor (SSRI). He began what became a series of trials of various SSRIs, all titrated in a very prudent manner. The patient became worse with each SSRI tried. Her agitation grew, not just initially, but in an enduring way. Her pacing and agitation increased. She began to have feelings of despair and transient thoughts of suicide. Her family became alarmed and she was referred to a psychiatrist. She was diagnosed with an agitated major depression. After failed trials with several other antidepressants from different classes, her suicidality had increased to where she was psychiatrically hospitalized. She received a course of electroconvulsive therapy (ECT) with minimal if any benefit.

At this point the family felt that the treating clinicians were out of ideas and asked if I would consult on the patient. The hospital graciously allowed me to do so and asked me if I would allow several psychiatric residents to accompany me. I interviewed the patients with the residents and asked them for their thoughts. Several suggestions for augmentation strategies were offered as well as the notion of a trial of a monoamine oxidase inhibitor antidepressant (MAOI). What I found incredible throughout this process is that no one seemed to have considered what struck me as the obvious solution for this woman's suffering. I recommended to the residents and subsequently to the attending psychiatrist that Ms W should be started on diazepam 2 mg three times a day. Initially there were objections to this idea. After all, the current literature sees SSRIs as the preferred treatment for depression and chronic generalized anxiety. Furthermore, the literature stresses concerns about the addictive risk of benzodiazepines as well as the risks of benzodiazepine use in older adults regarding cognitive clarity as well as the risk of falls. Now all of this is true. However, here is where having a model for the individual patient allows for the recognition of anomalies that suggest that an alternative treatment is best for this individual patient, regardless of what exists in the literature. I pointed out that diazepam not only worked initially for this woman but continued to work extremely effectively for over three decades. I also noted that there was no prior history of substance abuse for her or her family and that she had never engaged in any sort of drug-seeking behaviour, nor any interest in increasing her dose. Furthermore, her dose was low, there was no evidence of any cognitive impairment, at least not before the ECT, and no evidence of balance or stability problems. Thus, in her individual case the benefits of the diazepam vastly outweighed any potential risks.

The treating clinicians in this case appeared to be utilizing not a specific model for Ms W, but rather a more abstract notion of 'the anxious person', applying more generalized rules that would be appropriate in many cases but not for Ms W. In practice, a POP model should always be about a specific individual and their circumstances. The useful lesson here is that a clinician should always ask whether the model being employed incorporates unique features of this patient or is simply a more abstract picture of some theoretical type of patient.

I also suggested to the residents that Ms W's pharmacological response pattern, while an anomaly in relation to the most typical response of patients with generalized anxiety, was not totally unique. In my experience I have seen a small subset of patients whose generalized anxiety is worsened in an enduring way by antidepressants, especially serotonergic agents, even when carefully introduced. My general experience with these patients is that after multiple failed and painful trials of antidepressant medication, there is often an excellent response to long-term treatment with low to moderate doses of benzodiazepines. This illustrates how close attention to individual POP models can at times generate a new model of intermediate abstraction that characterizes a subgroup of patients. After this discussion the attending physician agreed to my plan. Ms W returned to her baseline functioning relatively free of anxiety within 2 to 3 weeks.

I am not suggesting that I did anything particularly clever in this case, although I must admit I was shocked that none of the other multiple clinicians involved in her care thought to do the obvious. I think the blindness came from the belief, perhaps unconsciously applied, that once the diagnosis was established the course of treatment followed based upon what is known in the literature. If instead one focuses on one's intuition about the specific pattern present in this individual, the anomalous response to treatment becomes glaringly evident. As is clear in this case, an anomaly need not be entirely unique to one individual. Often a result of accumulated experience carefully reflected upon is the recognition of subgroups of anomalies that become their own more fine-grained patterns that can help guide future treatment decisions.

9.8 Exposure to a range of patterns

What can be done to enhance training in the acquisition of such pattern recognition? This discussion will be framed in the context of formal psychiatric residency training, but it should be remembered that the principles involved apply to practitioners looking to enhance their expertise at all stages of their career. Klein's work stresses the importance of the trainee being exposed to a wide range of situations so that one may develop a good sense of both typical and anomalous patterns. Most current psychiatric residency programmes provide this. Residents typically rotate through a range of treatment settings, both inpatient and outpatient, and are likely to be involved in the care of a large number of patients throughout the training. In fact, this requirement can become more difficult for psychiatrists once their practices have become established in specific treatment niches. Psychiatrists in private practice must remember that a whole range of patterns exist that they may not have seen for a long time but may be common for psychiatrists working in mental health clinics or inpatient units. The converse of course is also true.

This is even a larger problem for the psychiatrist who practices in a very specific niches such as forensics, substance abuse, or academic settings. The academic psychiatrist in the current university climate faces a particularly difficult challenge. Academic success generally now requires focusing in a narrow specialty area and spending much of one's time doing formal research. Often a successful academic researcher will develop a reputation for an especially high level of expertise in a specific area and may often be called upon to consult by colleagues with particularly difficult cases. Often however these expert psychiatrists have been spending relatively little time personally responsible for the treatment of patients with the condition for which they are deemed an expert. It raises the question as to whether they are in fact the professionals in the best position to think through in a unique way the idiosyncratic needs of a specific and problematic patient, that is to develop an effective POP model for that patient.

9.9 Making models explicit: from goal to POP to simulation

Klein's work stresses that simple exposure to many cases is a necessary but not sufficient condition for the development of true expertise. Supervisors need to encourage residents to reflect upon the patterns they see. They can be encouraged to engage in and be guided in the process of making their hypothesized pattern for a given patient explicit, that is to articulate their hypothesized POP model. This should begin by stating as concretely as possible the goal of the model, what exactly is the patient wanting from treatment. The patient's goal is not always self-evident and initially may not correspond to the psychiatrist's initial assumption. The exact goal of treatment must ultimately be negotiated between patient and psychiatrist.

Part of what the psychiatrist can bring to the question is whether the goal is realistic. The psychiatrist must watch for one's own biases here as well as those of the patient. Psychiatrists may err easily in either direction. On the one hand, especially if we find ourselves liking the patient a great deal, we may want things for the patient that may be beyond that patient's capacity. Conversely, especially with chronically impaired patients, it is tempting to underestimate what is possible. Early in my career I had a particularly striking example of the latter problem.

Ms F and her family came to me because her psychiatrist of many years was retiring and she wished me to take over her care. She was 45 years old and married with three children. She had a history of four extremely chaotic psychotic episodes in her early 30s resulting in extended hospitalizations in each case. She was diagnosed with a schizophrenic disorder and treated with large doses of antipsychotic medication. These medicines did indeed resolve her psychotic episodes but left her essentially non-functional. She was sedated and very flat with prominent negative symptoms. She spent her days sitting and watching television with little interaction with her husband and children. Her husband, in addition to his career, provided the primary parenting for the children who at this point were in middle school. Her psychiatrist had told her that she had an extremely severe psychiatric illness, and while he understood the negative impact of the heavy medication, she and the family needed to understand that this was the best that could be done. The alternative would lead to a further stream of psychotic episodes and hospitalizations. In fact, she had not been hospitalized for over a decade at the time I saw her.

In taking her history, I felt there were some suggestions that she may be suffering from a bipolar I disorder rather than schizophrenia, and that her hospitalizations were in fact psychotic manic episodes. After extensive discussion with the patient and her family about the potential risks, I gradually tapered her off antipsychotic medication and began her on lithium. The lithium proved strikingly effective, and she remained on it for the next 30 years, until her death, without a relapse. Even more striking is how she came to life off all antipsychotics. She was able to reengage with her family and become a loving wife and mother. She started a small business that she was able to run successfully.

In this case I knew the previous psychiatrist and viewed him as quite competent. However, having witnessed her extraordinarily chaotic and disruptive psychotic episodes and finally achieving control of them, he reached the conclusion that this was as good as it could get. At that point I suspect he had stopped thinking of her as a unique individual understandable in terms of a specific POP model, and she became an instance of a category of chronic psychotic patient. I do not think I was especially clever here, but I was able to look at her with fresh eyes and consider whether a different goal was possible for her. Not knowing her initially, I felt the obligation to develop an understanding of her story, including her unique strengths and weaknesses, that could be incorporated into a POP model. Doing so made it easier to see the potential for new goals for growth beyond simply maintaining stability. This case has led me to make a conscious effort to look closely at my more chronic patients in those cases where I, and often the patient, have become complacent and accept a stable but rather low level of functioning as the best we can do. At times I have found it useful to ask for a second opinion in these cases with someone whose judgement I can trust to get a fresh look. Not infrequently they are able to see possibilities for the patient and modifications in the treatment that I failed to see.

As part of residency training most residents take on for some time the management of long-term and chronic patients. Often these patients, particularly if very stable, fail to be the focus of precious supervision time. Supervisors should encourage residents to put some focus on these patients as well. One strategy is to ask the resident to explicitly formulate a unique POP model for such a patient. This gets beyond the fatalism that frequently accompanies seeing the patient as simply an instance of a psychotic diagnosis, and forces the psychiatrist to look for the specific and unique features and circumstances of this patient, including strengths. In doing so it is common to discover some new potential avenues for treatment and possibilities for growth. Furthermore, residents need to learn to look repeatedly at the goals for all their patients. As several of the detailed cases considered in earlier chapters demonstrate, goals can change for a patient given new circumstances and responses to treatment. A new goal requires the construction of a modified or at times completely new POP model.

Once a goal is clearly articulated then the POP that the resident has hypothesized can be examined along with the range of interventions, or, in Klein's language the action sequence, can be looked at. Does the POP make sense in terms of the patient's history and our generalized knowledge base? Are there important factors that do not fit or that argue against the model? Do the facts in this case suggest an alternative model that might be more credible? Do the proposed interventions flow reasonably from the model's understanding of the patient and are they feasible in the patient's current context? None of the answers to these questions are infallible. However, articulating them consciously and explicitly allows obviously poor reasoning to be identified.

9.10 The need for prompt and clear feedback: Making concrete predictions

In addition, by articulating the model it should be possible to make concrete predictions of what should happen if the model is correct. Klein's work emphasizes that expertise can only develop with clear feedback as to outcomes, otherwise one can simply get locked into one's pre-existing biases as to what will work with a given pattern. This need for feedback poses several significant challenges in psychiatric training and in the practice of psychiatry generally.

9.10.1 The challenge of getting feedback in short-term treatment settings

In current treatment environments, residents, as well as established psychiatrists for that matter, frequently deal with patients for relatively brief periods of time, such as during an inpatient admission or a brief course of treatment in an outpatient clinic. As a result, a resident is frequently initiating treatment but loses contact with the patient before one has a clear sense of the impact of the treatment intervention on the course of the patient's difficulties. Hence there is no feedback as to the impact of the intervention. In this a critical learning opportunity is lost.

Early in my own psychiatric residency, on my first inpatient rotation, I treated a young woman who was admitted in a floridly psychotic state. Both I and my supervisor felt she had an illness on the schizophrenic spectrum. She was treated with antipsychotic medication and improved and was discharged. Shortly thereafter she was readmitted a second time, having stopped her medication, in a similar psychotic state. She was again treated with an antipsychotic, improved, and was discharged to a mental health clinic. Several months later I received a call from a psychiatrist in a nearby state psychiatric hospital. This patient was now admitted there. The psychiatrist initiated the call because she needed some specific information that I had about the patient and the family history. During our discussion, she pointed out that there were several features that could represent a bipolar disorder, and that she had placed the patient on lithium to which the patient was showing an excellent response with much better tolerability than she had to the antipsychotics. This was in the era of rather lengthy hospitalizations in state facilities, and the psychiatrist kindly invited me to keep in contact and get updates on her progress. In all of this I learned a great deal about features I had missed in my assessment of the patient leading to a very different understanding of her difficulties and the best interventions for her. In this case it was simply fortuitous that I was able to get this follow-up, and it contributed greatly to my understanding of this patient specifically, and more generally of the overlapping patterns of schizophrenic illness and bipolar disorder. Phrased differently, I had a tacit POP model for this patient, but the inadequacy of that model could only be assessed once I received concrete feedback of her response to subsequent treatment. One useful suggestion for residency training programmes would be to institute some mechanisms to allow residents to routinely get ongoing follow-up information and feedback on at least some subset of patients they have treated.

9.10.2 The challenge of precise prediction in long-term therapy

A second kind of problem in assuring prompt and precise feedback, rather the opposite of the one we have just considered, occurs in the context of training in long-term psychotherapy. Here the problem is not that of having adequate time to observe the patient, rather it is that often the goal of treatment is framed in rather global or abstract terms and may be relatively far into the future. Hence it can be a very long time, if ever, before there is clear feedback as to the effectiveness of the model and resulting treatment plan with this given patient. I suspect that the recognition of this difficulty has contributed to the common preference for more concrete and short-term behavioural and cognitive-behavioural therapies. These latter therapies clearly have their place and have added greatly to the range of options available in treating patients. Nonetheless, most clinicians continue to see a need for longer-term psychodynamic or interpersonal based therapeutic interventions for some patients.

How can we get meaningful feedback in these settings? Obviously the more specific and well-articulated the ultimate goal the better. A considerable lag time may transpire before we can assess whether our work is on the right track with this kind of treatment. Here it may be useful to make the model somewhat more fine-grained, even down to the level of a specific interpretation, from which a more short-term outcome can be predicted. When we make what we think is an important interpretation to the patient, what do we expect to happen if we are correct? Critics of these kinds of therapies frequently point out a dilemma here. If something gets better in a way we have expected, we take the interpretation to be correct. If the patient reacts negatively or gets worse, it is easy to maintain that the interpretation is correct but the patient's resistance and defences triggered a negative reaction. It is important here to remember, however, that a model's adequacy is not dependent just on whether the explanatory propensities it posits are correct, but whether the interventions we take as flowing from those are effective. If we make an interpretation that we hope will have a positive impact but it triggers a negative one, then the model fails in helping us find the desirable intervention at this point in the treatment. If the model did not allow us to understand that the patient was not ready to hear the interpretation at this point, it has fallen short of what it needs to do. We need to reassess the model, modify it, or replace it with a better one. Clinical humility, a virtue that I have argued tends to be promoted by the conscious use of POP models, once more comes into play here. An example may serve to illustrate this point.

Ms V was an unmarried nurse in her mid-30s. She was referred to me for pharmacotherapy by an experienced female psychodynamic psychotherapist who had been working with her for several years. The therapy work had focused on a history of extensive sexual abuse by her father during preadolescence. The therapist indicated that Ms V had been very guarded and only gradually was able to share some details of the sexual abuse with her. The therapist felt that the work had been progressing well if slowly, but the treatment was disrupted repeatedly by discrete bouts of quite severe major depression that made it difficult for the patient to engage in treatment effectively and also interfered with her day-to-day functioning. The patient had resisted for some time the therapist's recommendation of seeking psychiatric consultation for possible medication for her depression, but finally agreed. A condition of her willingness to see me was that I not probe into the details of her sexual abuse and she forbade the therapist to share any details with me. I had collaborated with this therapist in several cases previously, and she assured the patient that I could approach her sensitively and with minimal intrusiveness. Hence in starting with this patient I proceeded quite gingerly. She said that she was rather resistant to medication as she saw it as surrendering a degree of control, and that control was very important to her. After much discussion she agreed to take an antidepressant and was begun on fluoxetine. In taking the history she vaguely acknowledged the sexual abuse but did not feel comfortable discussing any details with me. I accepted this and told her that if she wished to discuss it with me at any time she could. She showed an excellent response to medication and the symptoms of her depression lifted.

I continued to follow her for brief medication visits over the next number of months. Although I felt we had a workable if tentative therapeutic alliance she became increasingly hostile and rather nasty towards me as time went on, even though she remained free of clinical depression. She accused me of being insensitive, uncaring, and uninterested in her feelings. She felt that all I wanted to do was 'shove medicine down her throat'. Shortly thereafter she stopped her medication, experienced a severe depressive relapse, reluctantly returned to treatment, restarted medication, and experienced a resolution of her depression. Each time her depression resolved, she acted as if I had in some way defeated her and was gloating. This pattern was repeated multiple times over the next couple of years. Throughout this I discussed matters with her therapist. Ms V had shared the same impressions of me with her therapist. Since we had shared multiple cases before, the therapist had a reasonably clear sense of how I practice and interact with patients. She reflected to the patient that this view of me was totally at odds with any feedback she had heard from other patients. It was also striking that for all her verbal attacks she kept coming back after each relapse rather than changing psychiatrists. Several previous attempts to work with psychiatrists had led to her rapidly stopping treatment.

Thinking of this case in terms of POP models, the model I initially had needed to be replaced. The initial model had to do with how to effectively treat her depression being aware of the larger context of sexual abuse. In essence the model worked as her depression was effectively treated. But now the goal of treatment, the problem to be addressed, was different. We now needed a model to understand her non-adherence and how to address it to avoid the repeated depressive relapses that occurred whenever she stopped her medication. The model had to explain her reaction to me and how it led to her non-adherence. Furthermore, it had to suggest some intervention to remedy this.

The model I developed hypothesized that I and previous treating psychiatrists, all male, were experienced as malignant male authority figures in the pattern of her father. Her repeated phrase that all I wanted to do was 'shove pills down her throat' led me to suspect that the abuse took the form of forcing her to perform oral sex on him. I further hypothesized that despite her overt hostility towards me, the fact that she continued to come back suggested that there was some positive alliance that might allow her to tolerate a challenging interpretation. Based on this, at our next meeting, when she was on medication and depression-free, and when she began her usual accusations towards me, I told her that I had been trying to understand her difficulties with me and that I wanted to suggest to her a possible explanation. I told her that although I had respected her wish not to discuss any details of her sexual abuse, I thought it was important at this point for me to address it. I told her my suspicion about the forced fellatio and that was up to her whether she confirmed, denied, or refused to discuss it. Further, I suggested that this might explain how she experienced the prescribing of medication from a male authority figure, and the resulting sense of loss of control and violation. I offered this idea for her to consider and indicated that we could discuss it now or in the future, but only if she wished. She remained quiet and thoughtful for a time and then acknowledged that I was right about the nature of the sexual abuse. She was not sure about the rest of it but that she would think about it and decide if she wanted to discuss it further. She never brought the matter up again with me. However, from that point on there was a dramatic change. The accusations stopped and she became friendly and appreciative of what I had done for her. She became totally compliant with her medication and remained depression-free for the next number of years until she moved to another part of the country for a career advancement, stopping treatment with me.

This case illustrates the need to revise models to address a new difficulty. Furthermore, it suggests how even in the context of a psychodynamically based understanding of a problem, a sufficiently fine-grained prediction can be made to allow meaningful feedback as to the adequacy of the model. Had she reacted in an extremely negative way and terminated treatment, which I realized was a real possibility, the model would have failed. It may have been correct in the understanding of the evolution of her attitude and her nonadherence. However, part of the model also included a propensity for a positive connection with me that I hypothesized was strong enough to tolerate the interpretation and that I could employ as an asset in seeking the desired outcome. Had the patient become enraged and terminated treatment, that part of the model would have failed. This contrasts with my experience with Ms M. In that case I hypothesized that the positive connection to me was strong enough to overcome her complicated transference and keep her productively in therapy. In that instance the prediction failed and a subsequent modified model was required.

9.11 Promoting clinical virtues: honesty, humility, and collegiality

The rigorous use early in training of POP models and the focus on concrete predictions from those models also serve to inculcate several important virtues in the psychiatrist. Foremost among these are honesty and humility. Psychiatrists must have the courage to honestly assess the adequacy of their clinical thinking based upon whether the predictions actually occur. They must also have the humility to accept the inadequacy of the models, and thus of their clinical thinking, when predictions fail and to make the necessary adjustments to the models.

All these strategies that I am suggesting for developing expertise in psychiatric residents can be employed not only with actual cases residents are treating, but with other difficult cases that supervisors can share from their own experiences or even with simulated cases that contain challenging elements. Thus, the resident can be forced to wrestle with patterns that exist but for which they do not yet have direct experience. Also, as noted already, these strategies for enhancing expertise continue to apply beyond residency training. Experienced psychiatrists can make similar efforts to consciously articulate their grasp of patterns with the patients they treat. This can be enhanced by discussions with colleagues in either formal or informal ongoing peer supervision. Our acquisition of expertise is never complete.

9.12 Leverage points and choke points

In Klein's model, the intuitive grasping of patterns is followed by an active mental simulation during which the intervention suggested by the model, the action sequence, is rehearsed imaginatively with an eye to places it can go wrong. This cognitive rehearsal is done before the intervention is implemented. With highly typical cases the simulation can be virtually instantaneous. In the presence of anomalies, it may require more effort and time. The two phases of intuitive pattern recognition and mental simulation clearly blur into one another. Some of the discussion about how to enhance pattern recognition included reflections on playing out possible treatment interventions. Nonetheless there are some specific suggestions that can be made toward enhancing the capacity for mental simulation in training settings. Approaches to this issue can be extrapolated from Klein's work with the RPD model of decision-making.

Klein's naturalistic studies of expert decision-making show that experts tend to refine their proposed action sequence by identifying especially critical steps that he terms 'leverage points' and 'choke points'.[2] A leverage point refers to a link in the causal chain, or in POP language, in the pattern of interactions between propensities, where an intervention effort is likely to yield an especially strong positive impact on outcome. A choke point is essentially a negative leverage point, that is, an especially critical link where a potential obstacle is likely to have a disproportionately disruptive effect on the success of any intervention derived from the model. The identification of a chokepoint requires an adjustment in intervention strategy that will either neutralize or circumvent the obstacle.

Klein found that with growing expertise an individual is increasingly able to identify useful leverage points and choke points and use that awareness to developing more effective intervention strategies. He further found that addressing leverage points and choke points often involves some modification of the overall goal being pursued.

> The decision-maker, now acting as a problem solver, can try to use experience to identify leverage points and construct a new course of action. In evaluating whether the action will work, the decision-maker can now try to improve it, or try to work with a different leverage point. Another approach is to try to modify the goals which may result in the identification of different leverage points. The evaluation process can alter the understanding of the situation itself.[3]

The case of Ms V can be conceptualized in terms of leverage points and choke points, as well as in terms of reconfiguring the goal. Ms V's original goal for me was alleviating the symptoms of depression, while leaving the management of her sexual trauma to her psychotherapist. In the short run that goal was achieved quite well. This initial POP model is represented in diagram form in Figure 9.1.

However, her repeated nonadherence with medication and the resultant relapses required a further elaboration of the POP model for Ms V. At the level of the goal, the focus became one of addressing nonadherence. From the vantage of leverage and choke points, her traumatic filtering of her experience of me was identified as a choke point. In thinking about how to deal with this roadblock, I had to revise my POP model to develop an appropriate intervention strategy for the chokepoint. This revised and expanded model is represented in Figure 9.2. In this case the identification and amelioration of the chokepoint was critical to achieving success in treating this patient. An elaboration and revision of the model and the goal were needed in this process.

9.13 Shifting emphasis in POP models

In some cases, changing circumstances can leave the model essentially intact but require a shift in emphasis based upon the identification of the currently relevant leverage and chokepoints. The case of Ms C illustrates this shift in emphasis.

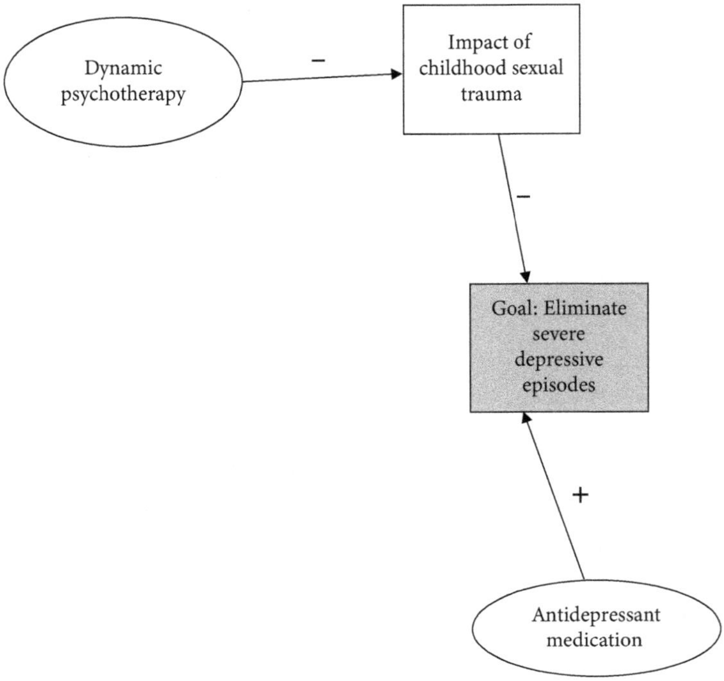

Figure 9.1 Initial model for Ms V.

Ms C at the time she first consulted me was a 45-year-old, married accountant with three teenage children. She and her husband were both professionally quite successful and worked long hours in their respective jobs. She described chronic depression of fluctuating intensity and significant generalized anxiety. She talked about how meeting the demands of work and raising her children were stressful for her. She felt that her husband offered little help with the children or the household. She reported that they had always had a poor emotional connection and she never felt that he met her emotional needs. Whenever she attempted to discuss this with him, he was unable to acknowledge any understanding of her issues and stated that he was completely happy with the marriage and that she should be as well. She periodically thought about leaving the marriage but was unwilling to consider it due to her children. She did have a positive family history for depression and anxiety.

This model proposed both a familial predisposition to depression and an important environmental contributor related to her stressed and overextended life in general and her problematic marriage in particular.

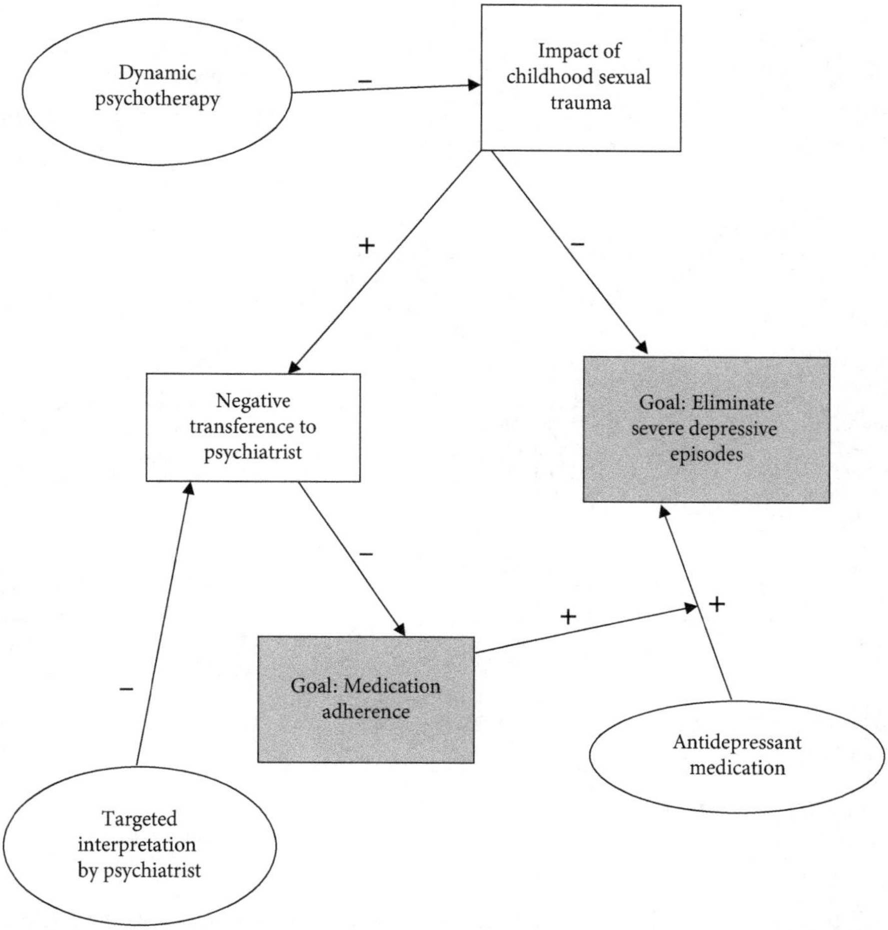

Figure 9.2 Revised model for Ms V.

Two interventions were proposed, namely antidepressant medication and psychotherapy. The latter initially included couples therapy, but this was quickly rejected by the husband. I continued to emphasize the importance of individual psychotherapy to address her troubled marriage, her stressful life, and to help her formulate a plan for her future. Although the patient was able to acknowledge in theory the relevance of psychotherapy, she resisted it saying that she already had too little time in her life and could not make room for regular therapy. She also said that she would not be willing to take any action as long as her children were still at home. As a result, I started medication and provided minimal supportive psychotherapy. Medication proved quite useful, reducing her depressive and anxious symptoms, increasing her energy and functional capacity, and decreasing her pessimism. Given what she was

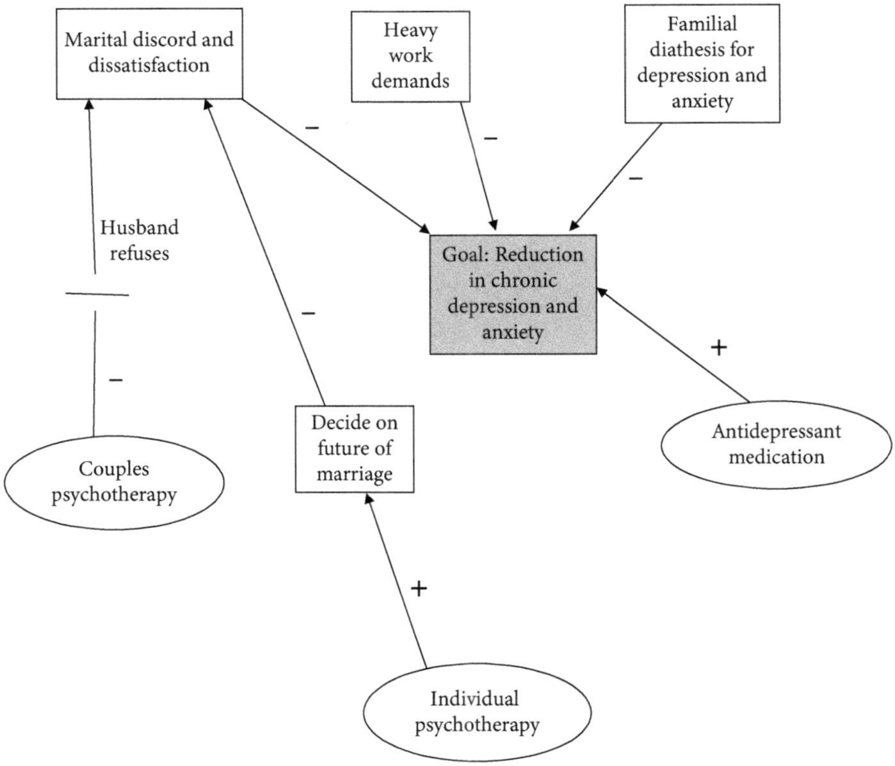

Figure 9.3 Model for Ms C.

willing and not willing to do at this point, the pharmacologic intervention became the primary leverage point. The POP model for Ms C is represented in Figure 9.3.

Nonetheless her general dissatisfaction with her marriage remained. She continued her antidepressants over the next five years. No further bouts of major depression recurred, though she remained chronically unhappy and dysthymic. When her youngest child began college, her depression increased significantly. She was now left alone with her husband, whom she found increasingly intolerable. She also was troubled by her own ageing and the realization that she might never achieve the emotional fulfilment she craved. She raised the question as to whether an increase or change in her antidepressants would be useful. I took the position that optimal benefit from antidepressants had probably been achieved. What was needed was to shift attention to the psychological work needed to decide about her marriage and to act on it. This issue of course was already captured in the original POP model. What has changed however is that it now became a leverage point that needed to be addressed if further improvement was to occur. As of this writing, Ms C has begun intensive individual psychotherapy and has separated from her husband, with a marked improvement in her mood.

As Klein observes, the expert's growing experience with a wide range of individual cases and their unique patterns enhances one's ability to identify and assess leverage points and choke points. That is clear in this case. One cannot go to the evidence-based literature to answer the question as to whether optimal benefit has been achieved with antidepressant medication for Ms C. Many other pharmacologic options are available, alone or in combination, that could be tried in the hopes of improving the response. Much time and energy would be consumed in this process. I would argue that most experienced psychiatrists, confronted with Ms C, would agree that that effort would be almost certainly futile, and that the leverage point that could make a real difference in her life would be to address her marriage and her own plans for her future life.

In considering how to enhance expertise among trainees as well as more experienced psychiatrists, educators should explicitly encourage clinicians to search for and identify leverage points and choke points in understanding the given patient's situation and in planning appropriate treatment responses.

9.14 Confirmation bias and the 'pre-mortem'

A second theme in Klein's work that suggests useful strategies to enhance the mental simulation of action plans is his emphasis on the fact that the leading pitfall for the expert is that 'once we have built a mental simulation, we tend to fall in love with it'.[4] As a result, the 'biggest danger in using mental simulation is that you can imagine any contradictory evidence away'.[5] This is an example of what Kahneman[6] calls 'confirmation bias', that is being more attuned to evidence that confirms our hypothesis and sidelining evidence that contradicts it.

> We can ask what evidence you would need to give up your explanation. Sadly, the answer is that if you are determined enough, you might never give it up. You can continually save the explanation by making it more comprehensive and complicated . . . People can also draw on some self-correcting characteristics of mental simulation, to overcome some of the limitations . . . It is often wise to explain away mild discrepancies since the evidence itself might not be trustworthy. However, there is a point when we have explained away so much that the mental simulation becomes very complicated. At this point, we begin to lose faith in the

> mental simulation and re-examine it. We look at all the new evidence that has been explained away to see if maybe there is not another simulation that makes more sense . . . Until we have an alternate mental simulation, we will keep patching the original one. We will not be motivated to assemble an alternate simulation until there is too much to be explained away. The strategy makes sense. The problem is that we lose track of how much contrary evidence we have explained away so the usual alarms do not go off . . . [What is needed is to develop] . . . training methods that will help people keep track of their thinking and become more aware of how much contrary evidence they have explained away so they can see when to start looking for alternate explanations or predictions.[7]

This would suggest that one important role for supervision would be to help the trainee to remain cognizant of contrary evidence for a model of the patient, and where appropriate, challenge one to consider alternative models that may do a better job.

One strategy proposed by Klein is the pre-mortem exercise.[8] One is asked to imagine that one has implemented the action plan proposed in the model but that it has failed. One is then asked to consider the most likely explanations for the failure. This essentially forces one to look for contradictory evidence and hence helps neutralize the confirmation bias. How might one proceed to do this? One could first look for any obvious or likely chokepoints in the model. As a more thorough exercise, one could examine each propensity proposed in the model, as well as each pathway of interaction between propensities, and ask what is the nature and strength of the evidence for that specific piece of the model. Or asked another way, how strong is the probability that the role attributed to a given propensity in the model of this patient is valid?

9.15 Towards a taxonomy of evidence

This discussion requires a more comprehensive concept of evidence than that usually associated with EBM as well as a notion of probability that goes beyond the simple statistical probability that forms the unquestioned backbone of much current academic research in psychiatry and medicine. Mathematical formulations of probability only began to appear in the seventeenth century in the context of calculating odds in games of chance, and only later applied to other human activities and questions.[9] This mathematical probability only applies fully in situations where there is uncertainty in outcomes but no ambiguity in how data is generated. Games of chance are the clearest illustration. How data is generated from the rolling of dice or the distribution of cards is not ambiguous. In real life, however, much of our uncertainty comes from the extensive ambiguity of real-life situations. Here the value of pure statistical analysis is limited because of the myriad variables at play. Other less easily quantified factors come under consideration requiring a much broader concept of probability and likelihood.

This broader concept of probability and of evidence has a much longer intellectual history. In our modern society these two views of probability live side-by-side often in relative disregard of one another. One place where the older notion of probability is explicit is in our legal system. Concepts such as probable cause, beyond a reasonable doubt, or the preponderance of the evidence are not reducible to mathematical quantities. They refer to comprehensive judgements of a wide range of types of evidence in the light of considerable ambiguity. This is closer to the concept of the strength of the evidence and the probability of its validity in our current discussion.

To keep this process from being simply vague and impressionistic, it is useful to articulate some of the common types of evidence and the bases for our judgements as to their strengths and weaknesses. Hence in considering each propensity and interaction in a patient model, we can interrogate the kind and quality of the evidence we are using. This requires a taxonomy of types of evidence that can be used in clinical thinking. In this preliminary discussion, I will suggest several common sources of evidence, assessing their features and relative strengths and weaknesses. There are likely others types of evidence as well, and their elaboration would be a useful part of a future pedagogical research programme in the development of psychiatric expertise.

9.15.1 Evidence from formal clinical research

One obvious type of evidence is that derived from formal research studies, either controlled trials or naturalistic studies of treatment or course of illness. Although the idea for any study may be inspired by a particular theoretical viewpoint, the evidentiary value of the study derives primarily from the empirical results. This sort of evidence is the most highly prized by EBM, and its strengths and weaknesses are those we discussed earlier when we considered EBM in more detail in Chapter 2. Its strengths derive from the ability of the study designs to control for variables and certain types of internal bias, and the ability to subject the results to various kinds of statistical analysis. Studies of this sort also have several kinds of weaknesses. For many questions of interest, good quality studies are few or non-existent. Not infrequently, study results conflict. Often effect sizes[10] are modest at best. External validity poses a large problem, that is how much confidence can we have that the study results are applicable to the specific patient for whom we are designing a model. Formal studies often have limited generalizability to clinical practice. A shared diagnosis between the individual patient with whom the psychiatrist is working and clinical trial participants can be a weak support for the applicability of the trial to the given patient.

Nonetheless, in many situations where we lacked good theoretical understanding of mechanisms underlying the condition both for our patient and for patients in general, a diagnosis provides a stopgap that allows us to posit one or several propensities and interactions that contribute to our model for the patient. Thus, in the case of Ms K, the judgement that she shared features of generalized anxiety disorder, even though she did not technically meet the full criteria for the diagnosis, led to the postulation of propensities and their interactions as parts of the model for understanding this patient and generating hypothesized interventions even in the absence of any clear understanding of the mechanism for generalized anxiety disorder (GAD) in this patient or for the group of patients who have this diagnosis. My use of 'stopgap' here addresses the hope that at some point we will have a better understanding of the mechanisms involved.

9.15.2 Hypothesized mechanisms as evidence

A second type of evidence is involved when we postulate certain propensities and their interactions based upon some theory about the underlying mechanism. Typically, these theories are derived from work in the closely related supporting disciplines for psychiatry, such as neuroscience, biochemistry, genetics, psychology, and the social sciences. The strength of this type of evidence is that it is embedded in a hypothesized understanding of underlying mechanisms. Of course, the strength of the evidence supporting these mechanisms varies immensely. Furthermore, the clinician is often not in a terribly strong position to judge how good the evidence is, as the work generally comes from a field other than the clinician's own area of expertise. Beyond this is the problem of translational understanding; namely, even if the theory is correct, how it translates into clinical presentation and interventions is usually highly speculative. The complexity of factors both within the individual, biological and psychological, and in the environment are so numerous that this direct translation is fraught with peril.

It is interesting to consider in biological psychiatry for instance, how many proposed treatments based upon such theoretical speculation have proven to be disappointing or failures.[11] Conversely it remains a fact that most of the major therapeutic breakthroughs in psychiatry have come from serendipity and astute observation of unexpected effects rather than from any theoretical understanding.[12]

9.15.3 Unstructured clinical observation as evidence

A third type of evidence, unstructured clinical observation, is that derived from recurrent patterns that the individual clinician has observed in treating other patients who resemble the patient in question in important ways. This kind of evidence has generally been strenuously devalued in recent years, particularly by advocates of EBM. There is no question that these kinds of observations have significant potential for bias and often draw on a small and perhaps not representative patient sample. They certainly should not be accepted uncritically by the clinician. Nonetheless, as we have seen, Klein's work on the development of expertise strongly argues that there is real and valuable knowledge that can accrue here. Its strength is that it derives from detailed exploration of individuals patients in their complexity. Hence this sort of data carries some evidentiary weight and should not be dismissed out of hand. Hypotheses derived from this sort of evidence are generally stronger when the expert has had good access to feedback on patient outcomes and especially when that feedback is reasonably prompt after the intervention. Hence a clinician's sense of pattern is likely to be more accurate in predicting short-term consequences than long-term ones. In a perfect world, formal clinical studies would be most helpful focusing on longer-term questions of outcome and course of illness where the clinician's own impressions are least helpful. Unfortunately, for a range of pragmatic reasons, a high percentage of formal studies focus on short-term outcomes. Hence psychiatrists should certainly take their own experience patterns seriously but tentatively, especially in the absence of other forms of supporting evidence for a particular propensity or interaction. My comfort in recommending that Ms W (section 9.7) be returned to low dose diazepam was strengthened by having seen a small group of anxious patients who shared her poor reaction to SSRIs and an excellent and unproblematic long-term response to modest doses of benzodiazepines.

An extension of the clinician's unstructured clinical observation comes from sharing in the communal experience of clinicians in the field. A vast array of clinical 'pearls of wisdom' shared over generations in the training of psychiatrists has been passed on by supervisors, teachers, and experienced colleagues. These can take the form of clinical maxims or other postulated generalizations about subgroups of patients or clinical situations that have not, at least not yet, been subject to formal study. As an example, 'When a suicidally depressed patient inexplicably becomes calmer and more at peace, it is important to explore whether they have reached a decision to commit suicide'. Or, 'When a psychotherapy patient begins to regularly forget appointments, this may be evidence of a resistance to treatment that needs to be addressed'. Or 'It is best to start a clinical evaluation of new patients by encouraging them to describe their reason for consulting the psychiatrist in an unstructured way and in their own terms before asking lots of specific questions'. Such shared observational evidence adds credibility to the unstructured clinical observations of an individual clinician. Thus, my confidence in the existence of that subgroup of patients that included Ms W was bolstered by the fact that several other experienced colleagues with whom I discussed this case shared similar observations from their own experience.

A more structured form for this sort of observation is the case report or case series. Again, this kind of evidence is far from infallible and subject to a range of biases. Proponents of EBM frequently focus on instances when well-established clinical beliefs proved to be incorrect, or at least are contradicted by the results of formal studies. Such instances clearly exist, and their correction is highly desirable for the field. On the other hand, more often these sorts of clinical beliefs turn out to be essentially correct or highly useful. It is easy to underestimate how often this is the case, as less controversial beliefs deriving from unstructured clinical observations are less likely to be subjected to clinical trials than controversial ones or those where some contrary evidence exists. After all, for a range of practical reasons, such a small percentage of important questions in psychiatry will ever be subjected to formal study. We should also remember how often new formal studies contradict not only some piece of accepted clinical wisdom but contradict prior formal studies as well. Neither are infallible sources of evidence, and both need to be regarded with appropriate scepticism by the clinician. But neither should be dismissed as having no evidentiary value.

Historically, this sort of unstructured observation has been a rich source of hypotheses for more formal research studies. Healy[13] describes how in the early days of antidepressant development, investigators typically administered a compound of potential interest to a wide range of patients on a psychiatric unit and observed the range of effects, describing them in detail.

From these unstructured observations, a more specific hypothesis about the use of the compound for a certain sort of patient could sometimes be generated and then subjected to structured clinical trials. As a more circumscribed example, it was initial unstructured observations for the anti-smoking properties of bupropion that led to confirming clinical trials and the marketing of bupropion for smoking cessation.[14]

9.15.4 Folk psychology as evidence

Finally, I want to address another kind of evidence that can best be characterized as 'folk psychology'. Hutto and Ravenscroft[15] identify several senses in which this term is used. In the broadest sense it refers to a set of human cognitive capacities that includes the ability to predict human behaviour in other members of our species in a wide range of situations using our capacity to attribute mental states to others and to explain their behaviour in terms of their possessing those mental states. Some students of folk psychology go further and argue that when we engage in this activity, essentially a form of mind-reading, we are applying a quite sophisticated if largely unconscious theory of human behaviour to which the term 'folk psychology' refers. This argument draws on the notion that many of the factors driving the evolution of cognition in humans related to the need to navigate the increasingly complex social world of our species at least as much is learning to manipulate the material world.[16] Hence the principles of folk psychology are actually the product of a long evolutionary process leading to the selection of beliefs about what drives the behaviour of other people that prove efficacious for our survival and success as members of the human community.

Folk psychology generally explains people's behaviour in terms such as their beliefs, their desires, and their expectations. We all employ this kind of thinking so extensively and unthinkingly that it is easy to miss that it relies on a sophisticated theory about how humans operate. As clinicians, a high per cent of our judgements and expectations about patients derived from folk psychology. If I say something to offend a patient who reacts angrily during the session and then fails to show up for the next scheduled visit, I do not need the theories of Freud, cognitive-behavioural therapy, or any other articulated psychological theory to draw the obvious connection. I quite automatically postulate that if the patient felt unfairly treated by me, that belief will lead to an emotion of anger and that the anger may well lead the patient to do something that they think is punishing me in return. When we build models that attempt to explain the patient's difficulties and to suggest useful interventions, a high percentage of the propensities and their interactions that we hypothesize come from folk psychology.

How seriously can we take this kind of evidentiary support for the components of our models? It is easy to dismiss folk psychology as something that is primitive and reflexive, and hence to assume that it is naïve to take it as a serious source of knowledge. It certainly can be fallible. It is also true that two individuals might well interpret a specific situation differently in trying to understand the behaviour of another although they are both employing folk psychology. Our own unique experiences and situation can certainly modify our specific reasoning about the behaviour of another. Judgements can also be affected by differences in cultural, racial, or socioeconomic factors. Nonetheless it is undeniable that as a species we have developed highly successful capacities to predict the behaviour of others, capacities that have been essential for the development of complex human societies. Folk psychology has developed over a long period of time under powerful evolutionary pressures to be generally accurate most of the time. It constitutes a coherent network of heuristics. Hence, we should take it seriously. One of its potential weaknesses of course is that it generally operates at a largely unconscious level. As a result, any biased or irrational aspects of our thinking can remain unexamined and uncorrected. One way to attempt to compensate for this is to try to make our exercise in folk psychological theorizing as explicit as possible when we are employing it in generating clinical judgements. The concept of folk psychology is one way to understand the distinction of the philosopher/psychiatrist Karl Jaspers between *erklären* and *verstehen*,[17] discussed further in section 7.11. The latter refers to psychiatric knowledge not based upon collected empirical data, but deriving from our capacity to find the patient's behaviour in a specific situation as understandable and meaningful in human terms. This can be thought of as a capacity for empathy, but this way of thinking can sometimes lead to seeing it as a mysterious or spiritual attribute. I think it is best to view it as simply a situation that makes sense in terms of our underlying folk psychology, in terms of how we understand human beings to act given their beliefs, their desires, and their emotions.

We should note that many of our ways of thinking about patients that are derived from psychologically based theories really represent elaborations and refinements of folk psychology. So the basic tenets of the cognitive-behavioural theory of depression, which argue that holding negative beliefs and expectations about oneself and one's situation can be a driving force in creating and maintaining depression in a patient, is not that removed from folk psychological understanding that self-deprecating beliefs can lead to negative feelings about oneself and one's possibilities in the world and result in one's giving up on pursuing precisely those goals that would make one feel better about oneself. Similarly, many features of psychoanalytic theory, especially what is regarded as clinical theory, represent elaborations of folk psychology. The characterization of repression and of classic defence mechanisms is a case in point. One needs only look at a wide range of world literature before Freud to see instances of these mechanisms that are clearly understood by the authors and the readers based on our shared notion of how humans operate. We do not need Freud to find Lady Macbeth's handwashing to be a response to her guilt. What this aspect of psychoanalytic theory achieved was to make these mechanisms more systematically understood and to emphasize something that indeed is already understood in folk psychology, namely that much that drives human behaviour is done at a level of which the individual oneself is unaware but that still can be understood in terms of concepts like beliefs, desires, emotions, and expectations.

9.16 The centrality of models for developing psychiatric expertise

What I have been suggesting is that one way to help develop the expertise of psychiatrists, and to improve the accuracy and credibility of their pattern recognition and mental simulations of POP models, would be to engage in the exercise of asking what kind of evidence we have for each proposed propensity and interaction. Then one can consider the strengths and potential weaknesses of that kind of evidence. This allows a thoughtful premortem exercise. If our model fails to fulfil our expectations, what are the weakest links to which we should look in our efforts to understand that failure and to develop a better model in its stead?

In this chapter I have used the insights into human decision-making and expertise that have come from careful research in those areas to suggest some avenues by which POP modelling can be used in training to enhance the development of expertise in young psychiatrists. I believe the same avenues can be used throughout the psychiatrist's career, on one's own or with peer collaboration and supervision, to continue to develop higher levels of expertise over time. These suggestions have been very general but I hope not vacuous. I believe this area of pedagogy has been relatively undeveloped in any explicit way. Rather than these suggestions being a finished product, I hope they point towards a pedagogical research programme in psychiatric training to develop and refine more explicit strategies for enhancing clinical expertise.

We have thus far characterized POP model construction in psychiatry. We examined how it functions as a descriptive framework to articulate how good psychiatry is in fact practised. We next looked at some practical or instrumental uses of POP models in psychiatric training and for the enhancement of expertise in the practice of psychiatry generally. In the next chapter we will take up the issue of the third function of the POP model, namely the use of this framework as a basis for justifying the reasonable knowledge claims embedded in practising psychiatrists' hypotheses about their understanding of specific patients—their difficulties and predictions about their responses to proposed interventions.

Endnotes

1. Giere 1999, pp. 97–117
2. Klein 1998, pp. 115–123
3. Klein 1998, p. 118
4. Klein 1998, p. 71
5. Klein 1998, p. 68
6. Kahneman 2011, pp. 80–81
7. Klein 1998, 69, pp. 72–73
8. Klein 1998, pp. 74–75
9. Debnath and Basu 2015; Hacking 2006; Weisberg 2014
10. In treatment trials, effect size refers to the magnitude of the difference that the experimental intervention makes compared to the absence of the intervention on some measured outcome. This contrasts with statistical significance (the p value) which refers to the likelihood that the observed difference is a chance occurrence. Especially with large sample sizes, a treatment effect can be highly significant but have a very small magnitude of effect. Effect size is sometimes called a measure of clinical as opposed to statistical significance. Depending on the design of the study, effect size can be characterized as the difference in outcome between groups or the strength of the association between variables. Indices used to quantify the former include Cohen's d, odds ratio (OR), and relative risk (RR). Indices for the latter include Pearson's r correlation and r2 coefficient of determination. For an excellent and concise overview of these issues for the non-statistician, see Sullivan and Feinn 2012
11. For instance, see Kendler and Schaffner 2011
12. For instance, see Healy 1997
13. Healy 1997
14. Tong *et al.* 2006
15. Hutto and Ravenscroft 2021
16. Cosmides 1989; Mercer and Sperber 2017
17. Jaspers [1913] 1997

Chapter 10
Justifying knowledge claims in psychiatric reasoning

10.1 Epistemological justification in the scientific age

In our journey thus far, we first argued that evidenced-based medicine (EBM) does not provide a comprehensive framework for clinical reasoning and decision-making in psychiatry proper. We then developed an alternative understanding of what a good clinician does in reasoning about an individual patient, using a series of case studies. The result was the articulation of the pattern of propensities model (POP) by which the patient's difficulties are formulated as a pattern of propensities and their pathways of interaction that become potential targets for therapeutic interventions. The POP model is intended to describe explicitly what good clinicians in fact do in making decisions about treating patients.

We then examined some of the potential practical gains from the explicit use of POP models for the development of expertise for the practising psychiatrist. These pedagogical approaches have clear relevance to the training of psychiatrists but also apply to the career-long efforts of psychiatrists to improve their expertise. As background for these practical approaches, we reviewed some of the major frameworks for thinking about human reasoning and decision-making, culminating in a focus on the work of Gary Klein and others who have explicitly studied the strategies used by experts in a range of fields. With growing experience and expertise their mode of reasoning relies more and more on heuristics of intuitive pattern recognition and mental simulation. These processes, although described in somewhat different language, essentially constitute the POP model. Having developed my notion of POP modelling over many years, and before I discovered Klein's work, I find this convergence very reassuring.

Having discussed the descriptive and the pedagogical functions of the POP model, we now turn to a third function, that of the epistemological justification of clinical reasoning. On what basis we can have confidence in the knowledge we assert to know about individual patients based on POP models? Modern medicine, including psychiatry, has been increasingly

How Psychiatrists Make Decisions. Douglas W. Heinrichs, Oxford University Press. © Oxford University Press (2025).
DOI: 10.1093/9780198960720.003.0010

committed to the notion that what we claim to know should be based on solid reasoning and evidence and not just on opinion, guesswork, traditional medical lore, or habits of practice. One of the primary drivers of the rise of EBM has been that it claims to represent a truly scientific approach to medical practice and that hence the conclusions we draw for an individual patient based on EBM methods represents the best of scientific practice.

It is worth noting here that the question of epistemological justification becomes essentially synonymous with arguing that what one is doing is scientific. It is easy to understand why this would be the case. After all, our modern age is one in which the tremendous advances in all fields of science and technology have led to the belief that scientific method provides our strongest justification for accepting assertions about the world as credible and worthy of belief. As we have seen in Chapter 2, the problem with EBM is that applying results of clinical trials on aggregated patients to the individual patient is fraught with problems and certainly cannot be taken at face value as scientifically valid.

What then are we to make of the epistemological status of the models we develop when employing POP reasoning in treating the individual patient? This is a fair and important question. Articulating an answer to this question will be the focus of this chapter. It will require a somewhat lengthy detour into notions elaborated by the philosophy of science over the past 60 years, but I believe that the reader will find this a worthwhile and relevant journey.

There is a reason such a detour is needed. In our current world, the most convincing way to establish epistemic credibility for any intellectual activity is to argue that it is an instance of the scientific method at work. This in turn assumes that we share a clear understanding of what constitutes scientific method. As with so many of our most pervasive concepts, it is easy to assume we have a clear understanding of what constitutes scientific method, but on closer examination there has been and remains significant ambiguity and controversy. The philosophy of science has been an area of active and explicit inquiry for well over 100 years, and less explicitly as a part of epistemology for much longer. Fitting with the philosophic examination of any discipline, it looks to make explicit and critique the underlying assumptions that guide the practice of science.

10.2 Three epistemological frameworks

We will consider three possible positions offered as to the relationship between scientific method and the method of clinical reasoning about the individual patient that we have made explicit as POP modelling. As we do so, we need to ask ourselves what is the understanding of scientific method underlying each of these positions. The first, which coincides with the position of most adherents of EBM and their sympathizers, is that the scientific status of a clinical judgement is an appropriate criterion for its trustworthiness, but that clinical reasoning about the individual patient has low epistemological status because it is not scientific. Hence, while the lack of more complete scientific knowledge in psychiatry may make this kind of clinical reasoning necessary in many situations, in an ideal world it would eventually be largely replaced by firm scientific knowledge based on empirically verified algorithms and valid diagnoses.

A second position would be to agree that clinical reasoning about an individual patient is not science but to maintain that a different criterion of trustworthiness should be developed and applied. Adherents of this position tend to stress that the understanding of a unique human being is such a radically different enterprise from science that scientific method does not provide the appropriate epistemic standard.

The third position has up to now been rarely taken, but it is the one I intend to defend. This position asserts that it is appropriate to apply standards of scientific method to the process we employ in understanding and making decisions about the treatment of an individual patient and, that if the standards of scientific method are properly understood, the methodology of POP modelling passes muster as a scientifically valid procedure.

The first two positions, despite their radical differences, share a similar view of what constitutes scientific method. I will argue that this view is grounded in an excessively narrow and now largely rejected characterization of a common method applying across all scientific disciplines. I will further argue that an alternative understanding of scientific method casts a very different light on the assessment of the scientific status of clinical reasoning about an individual patient using POP modelling.

10.2.1 Hermeneutics and narrative

Let us look at the second position first. Adherents of this stance argue that the task of understanding the human being is fundamentally different than the deterministic explanations of the rest of nature by means of scientific theories. They stress that the way we come to know about another human being occurs in the context of an evolving relationship between patient and clinician that requires an open-ended receptivity on the part of the clinician to whatever the patient brings to the encounter. The understanding of that patient, based largely on first-person and second-person perspectives, is something utterly unique and requires an empathy and receptivity quite unlike the detached, third-person stance that characterizes scientific exploration. The psychiatrist/philosopher Karl Jaspers argued that this led to a kind of understanding of patients' situations, *verstehen*, based on their motives and experiences being comprehensible to us as fellow human beings. He contrasted this with third-person knowledge, explanation, or *erklären*, based on a causal understanding of patients' behaviour. He argued for a role for both approaches in clinical practice, but saw their processes and products as quite distinct.[1]

In the view of advocates of this second position, reducing a patient to a set of causal hypotheses and explanations does violence to the integrity of the individual patient. Furthermore, the existential freedom of the individual person makes any attempt to apply a deterministic model inappropriate. The need to incorporate existential human freedom is a crucial consideration for advocates for this position. They frequently draw upon a rich tradition in the philosophy of the social sciences known as 'hermeneutics'.[2] Although originally referring to methods for the exegesis of biblical texts, its application has been extended to cover a wide range of theories of interpretation in general, at times conflicting with one another. I strongly believe that the overall understanding of interpretive strategies offered by the hermeneutic tradition provides a rich resource for the practising psychiatrist struggling to understand the full complexity of an individual patient, and these strategies deserve to have a prominent place in our work. However, there is one thread of the hermeneutic tradition that I believe can be a source of mischief in our efforts to develop an appropriately broad conception of scientific method.

Beginning in the 19th century, with thinkers such as Dilthey,[3] there developed a project under the name of hermeneutics to formulate an interpretive method appropriate to history and the social sciences that was explicitly contrasted with the methodology of the natural sciences. In the case of history, one is trying to understand singular events, such as the French Revolution, that are embedded in broad societal/cultural relationships. A form of thinking called the 'hermeneutic circle' or 'round' is a pivotal component of this methodological approach. In a general sense, the hermeneutic circle describes the process of interpretation as a going back and forth between partial understandings, where a partial understanding of the unique particulars of the situation informs a more generalized understanding of this type of situation, and conversely the generalized understanding returns to revise our understanding of the particulars, recursively. Hence one moves from an understanding derived from looking at the unique individual case to alter one's view of the more general issues which in turn helps one revise the understanding of the individual case. How can the understanding gained from this approach be best represented? Many adherents of this position argue that this is best done by narrative.[4] The product of this kind of understanding of the individual person is essentially a story about that person.

10.2.1.1 POP models facilitate relevant narrative: The hermeneutic circle

In appraising this position, I want to start by stressing total agreement with the view that the way we come to understand another person involves a methodology rooted in a relationship with that person characterized by openness, dialogical interaction, and empathy. This is something that I suspect very few psychiatrists would see differently. Indeed, for many of us part of the appeal of this medical specialty was our valuing that kind of relatedness with other human beings. The question is—what do we make of the information we obtain regarding other individuals over time and in the context of our relationships with them? What do we do with our story of Ms Jones? How does it lead to informed decision-making?

The first thing to note is that any story of Ms Jones that has any coherence and is not infinitely long must involve an active selection as to what is important and relevant to the clinical situation and what is not. In our talking with Ms Jones, we learn a vast array of facts, many of which we omit from the story because we believe they are not relevant to the task at hand. Our narrative of Ms Jones is highly selective. How do we make those selections and judge their salience for our task? Some framework of understanding must be imposed on the narrative. The POP model serves to articulate that understanding and make it explicit. One must identify factors that are judged to have some causal relevance to the concerns that bring Ms Jones to treatment and the potential interactions between those factors. It is by employing a POP model, implicitly or explicitly, that we weed through the plethora of individual revelations about Ms Jones that emerge in the context of our discussions and identify those factors that are most salient to our task. We are likely to leave out of our narrative things such as her favourite colour, what she had for breakfast yesterday, her mother's maiden name, and so on.

To identify the salient propensities and interactions using a POP model, the clinician must merge the individual factors in the patient's narrative with more generalized information derived from the empirical knowledge base of the field (knowledge about diagnostic categories, treatment studies, basic science understanding of relevant processes, etc.), shared clinical wisdom learned in training, and the clinician's own growing repertoire of models from the narratives of other patients one has treated. This involves a back-and-forth flow between the general and the specific that is totally compatible with a broad understanding of hermeneutics as an interpretive process across various domains of knowledge, and in fact is an instance of a hermeneutic circle.[3] From this back-and-forth movement between general knowledge and an individualized understanding of the patient, the clinician generates a hypothesized pattern of propensities relevant to explaining the patient's current situation and predicting the outcomes of certain interventions. It is this model in turn that helps to further identify features that are likely to be most salient to the task at hand. As I have already argued, POP models, as with any model, are always constructed from a particular perspective in relationship to a particular purpose or goal. As such there is always an interpretive dimension.

To illustrate how a POP model helps to identify salient factors in the patient's story, let us return to the case of Ms M from Chapter 6. My failure to retain her in psychotherapy led me to reconsider the adequacy of my model. In the original model I had put my primary emphasis relevant to the psychotherapy on the information regarding her relationship with her father. In contrast to the salience accorded this factor, I de-emphasized information relevant to her relationship to her mother, even though it was part of her initial narrative. In revising my model considering the treatment failure, I was led to speculate that the maternal relationship deserved much more salience than I had previously appreciated. This led to the decision to refer her to a female therapist, which proved highly successful. Models are not incompatible with hermeneutic understanding if they are not seen as deterministic, causal mechanisms. I specifically chose to characterize POP models as patterns of propensities so that they can be employed without having to take a stance on which elements are best understood as deterministic causes and which can be seen as acting on reasons, that is, as expressions of human freedom. The model is intended simply to aid in the prediction of behaviour and the impact of interventions. So, if I am attempting to predict the behaviour of an individual confronted by an armed man who states that he will shoot the individual unless he gives him the money in his wallet, I can make the prediction that the person will probably give over the money regardless of whether I see his behaviour as totally determined or as a reflection of a free and rational choice under the circumstances. Obviously that prediction may be wrong, but in most cases I suspect that I will be correct regardless of my metaphysical position on the question of free will and determinism. It is precisely to remain neutral on the question of determinism and freedom that the relevant factors in the model are viewed as dispositions or propensities rather than labelled as causes.

One of the strengths of an explicit model is that it forces the clinician to articulate the basis for selecting concrete interventions with Ms Jones. The choice of an intervention is based on a prediction of outcome, and the prediction of outcome is based upon the explanatory framework of the model as it links various dispositions and their interactions. While a narrative of the patient's story can be a very useful way to begin the development of an individualized understanding of a patient, a story does not in itself generate a prediction. There must be some inferential process that leads from the patient's story to a prediction about the outcome of a chosen intervention. The model serves to articulate this inferential process in a way that the story, by itself, does not. There is no intrinsic incompatibility between the use of narrative strategies to generate information and the use of models to formulate predictions and to make choices among potential interventions. Indeed, this is demonstrated in a recently reported Norwegian collaborative project between philosophers and clinicians, CauseHealth, that attempts to generate causal understandings of individuals' medical conditions from careful attention to their individual narratives.[5]

10.2.1.2 Must narrative be unscientific?

I maintain that POP modelling is totally compatible with a rich use of narratives and interpretive strategies derived from the hermeneutic tradition. On what basis, however, should we assume that this activity stands in contrast to scientific method? This is the stance of many self-identified advocates of narrative.

> A scientifically competent medicine alone cannot help a patient grapple with the loss of health or find meaning in suffering. Along with scientific ability, physicians need the ability to listen to the narratives of the patient, grasp and honor their meanings, and be moved to act on the patient's behalf. This is narrative competence, that is, the competence that human beings use to absorb, interpret, and respond to stories . . . Unlike its complement, logicoscientific knowledge, through which a detached and replaceable observer generates or comprehends replicable and generalizable notices [sic], narrative knowledge leads to local and particular understandings about one situation by one participant or observer . . . Logicoscientific knowledge attempts to illuminate the universally true by transcending the particular; narrative knowledge attempts to illuminate the universally true by revealing the particular.[6]

Advocates of narrative medicine certainly do not deny that the competent physician needs to integrate the two approaches. Nonetheless, as this quotation from one of the pioneer advocates of the narrative approach in medicine Rita Charon illustrates, they are viewed as different approaches. One is seen as science, and the other as something quite distinct from it.

What understanding of scientific method makes this opposition seem so fundamental and inevitable to advocates of narrative like Charon? The assumption would appear to be that science itself is committed to the premise that all entities in the world, including human beings, are elements in a vast and universal system in which each event is an instantiation of universal laws of nature being played out in a rule-determined way. In this perspective scientific explanation of an individual instance means identifying that instance as one embodiment of a class of similar instances explained by a general law or laws. As such, a true science of the individual that goes beyond reducing one to an instance of a general type is not possible. Is this the only way science can be understood? While it is true that at one time this was a fair characterization of how the scientific enterprise was viewed, we will shortly see that this is no longer generally the case.

I want to make a final observation about the second position before we move on. In taking the stance that our clinical understanding of patients involves something that is radically different from scientific method, one is conceding a great deal of potential status for the trustworthiness of clinical reasoning given the current value system of modern medicine, including psychiatry. As we noted at the start of this chapter, we live in an era where the presumption in favour of science as the privileged road to reliable knowledge is highly dominant.[7] This problem is reflected in the commonly used reference to the 'art' of medicine. As it is usually employed, this represents a very backhanded compliment. While it might be taken to suggest that the clinician has some sort of special knowledge not captured by science, when the chips are down and there is a conflict between a clinical judgement and what is perceived as the scientific knowledge base, the latter usually will carry the most weight. In modern medicine science generally wins out over art. Now if the second position were the only valid way to understand the process of clinical reasoning, then one should certainly stick to one's conviction regardless of the rhetorical and pragmatic consequences. I will be arguing however that it is neither the only nor the best way to conceptualize the process.

10.2.2 Law-based science: The received view

Before we discuss how best to conceptualize scientific method, let us look at the first possible position as to the relationship of clinical reasoning and scientific method. The adherents of this position, which includes the strong advocates of EBM, degrade the epistemological status of clinical judgement by seeing it as fundamentally unscientific. As we have seen, the judgement of the clinician about the individual patient clings to the very bottom of EBM's hierarchy of evidence. Advocates of the first position employ a similar concept of what constitutes the scientific enterprise as do those of the second position. The difference is that they see this approach as the best way to achieve trustworthy knowledge about the individual patient as opposed to rejecting it in favour of some alternative pathway to the truth.

The view of science implied by both positions sees the scientific enterprise as the search for generalized regularities in nature. The task for the clinician is to determine which general categories Ms Jones instantiates so that generalized findings about that group of patients can be applied to the treatment of Ms Jones. The diagnosis often serves as such a general category. Hence if Ms Jones meets the criteria for major depressive disorder, the treatment with the best support from controlled trials of groups of patients with major depression will be our prediction of the best treatment for Ms Jones. Ms Jones becomes a 'token', and instance, of a general 'type', i.e. major depression. I believe that this approach characterized the thinking of the expert in bipolar disorder as described in section 6.4 who for a time treated Ms M.

We have already considered in depth in Chapter 2 the many practical problems of applying this kind of method to an individual patient in making a treatment decision. What I wish to emphasize here is the view of science itself that is assumed by this position. Knowledge is seen to reside in generalizations about groups of individuals that have been obtained by inductive reasoning based on studies of sample groups of patients. Once those generalizations are in place, the results are deductively applied to Ms Jones by virtue of her membership in that class of patients.[8] We should note that from this viewpoint a science of a particular individual that intends to capture something of the uniqueness of that individual *qua* individual, 'Jonesology' in the case of Ms Jones, is an incoherent idea. In this view scientific understanding involves using a generalization to explain an individual instance and to make predictions based on that explanation. The generalizations, in turn, are not knowable *a priori* as with mathematical axioms. Rather they are discovered by induction using multiple observations from many instances. In ordinary life, this can be the inductive basis of predicting 'the sun will rise in the East today'. We have observed the sun rising in the East countless times in our lives, and today's sunrise is simply one instantiation of that generalization. We need no specific knowledge about today's sunrise to make that prediction.

In some scientific cases, especially in fields like physics, the resultant generalizations can be held as applying to a broad range of entities with tremendous consistency, e.g. gravitation or the laws of motion in classical mechanics. In other fields, including biology and medicine as well as psychology and social sciences, generalizations tend to be probabilistic and apply to much narrower groups of entities. The goal of a clinical trial is to provide a systematically collected set of instances for the construction of such a generalization. Regardless, this view of science, so dominant in the first two-thirds of the 20th century that it has been termed the 'received view',[9] does allow for defining narrower and narrower subgroups to which Ms Jones belongs, but a truly unique understanding of Ms Jones as an individual, beyond being a token of a particular type, could not be scientific.

10.2.2.1 The emergence of the received view

How accurate is this as a picture of what scientific method and practice really is? In short, is a science of 'Jonesology' possible? Perspectives on what constitutes scientific method have changed radically from the early and middle part of the twentieth century to the present day. It is time now to explicitly examine some competing views of what constitutes science and its assumptions and methods. Some historical perspective is extremely helpful in understanding how we think about scientific method and how we have come to conceptualize scientific knowledge.[10]

Very rapid progress in the understanding of aspects of the natural world were made starting in the seventeenth century that were deeply impressive to contemporaries as well as those who followed, indeed up to the present day. The world-altering application of scientific knowledge to the exploding technology of the industrial revolution and all that has followed to create our modern world further added to the status of science.

Several features of seventeenth-century science[11] left a deep influence on our thinking about what constitutes a legitimate scientific enterprise. First, the earliest and most dramatic achievements were in physics and the closely related field of astronomy, culminating in the achievements of Isaac Newton.[12] Newton's approach embodies several features that have been associated with the proper practice of science ever since. First, observation and experimentation were closely linked to mathematics. Great predictive power was achieved by sophisticated quantification of hypotheses and observations. Second, Newton saw his task as discovering universal generalizations about how all aspects of the universe behaved that could be understood as 'laws of nature'.[13] It follows that scientific explanations show that a specific state of affairs fits into a general category of events that follow a law of nature. What we have here is a deductive process, where one goes from a general and universal statement to say something about a particular entity or event.

The general law itself is not discovered deductively or with *a priori* reasoning. Rather general laws are the result of inductive inferences based upon making many observations of particular instances. Third, there is the assumption of the uniformity of nature. The same laws should apply wherever one looks, be it in the heavens and the movement of celestial bodies or the mundane behaviour of objects on earth. This stands in contrast with the earlier Aristotelian view that a different set of regularities applied in the celestial realm as opposed to those that applied on earth.

Is this the only way that scientific understanding could have been formulated? Or are important aspects of this view the product of contingent historical accident? The philosopher of science Ronald Giere[14] has argued that the use of the notion of 'law of nature' is not the only or even optimal way to conceptualize scientific knowledge. Rather he argues that this notion follows from the strongly religious position of seventeenth-century scientists. An all-knowing and all-powerful deity would run the world in an orderly way. The created universe would of necessity follow perfectly the laws dictated by its creator. Newton himself was a deeply religious man, in some ways even eccentrically so, and saw what he was doing as discovering God's laws for the universe.[15] Over the centuries as science has become more secularized, it has essentially eliminated the lawgiver. Even religious scientists are not likely to use the concept of God in explaining their work. But we have anachronistically kept the notion of laws. In a theological framework it makes sense that the law explains the behaviour being observed. After all, the law is something God is imposing upon his universe and it therefore has a causal role in bringing about happenings in the world. However, without the lawgiver, a law of nature—'law' now being used metaphorically—is simply a description of regularities observed. As a result, it would seem to lose its explanatory force. It becomes the tautology of 'this is the way things behave because this is the way things behave'. While greatly weakened as an explanation, the idea of a generalization may still have some value in predicting the behaviour of a particular entity if it can be shown that the entity is an instance (often referred to as a 'token' in philosophical discussions) of a category of entities (or a 'type').

The adequacy of this type/token relationship gains strength as the entities under consideration are viewed in more narrow and abstract terms. For instance, the description of an entity in physics is likely to be reduced to a small number of abstract properties, such as mass, position, extension, and velocity. The vast range of other properties that make that entity more richly characterized and more unique have already been excluded from consideration. Hence it is easy to see the entity as an instance of a category of entities that are totally interchangeable with one another. It is also noteworthy that this much smaller group of properties are those that are most easily quantifiable. This goes some way to explain how the application of general laws as a scientific explanation of a specific instance seems to make most sense and have the most predictive power in a field like physics as opposed to other scientific disciplines.

10.2.2.2 Laws without a lawgiver: Causation and the problem of induction

In the absence of the lawgiver, how confident can we be in this way of thinking? Without the lawgiver there is no basis for asserting that the law, the generalization, explains rather than simply describes, the behaviour of the token instance. The problem here was astutely captured by the eighteenth-century philosopher David Hume.[16] Hume's approach to knowledge played a pivotal role in the evolving understanding of science up through the first half of the twentieth century. Hume's epistemology was radically empiricist in nature. He held that all that we could really know had to be derived from observations. He argued that causality itself is not observable, but is rather an inference that humans make when they consistently see event B follow event A.

To use an example from Hume, when we see a billiard ball rolling across the table and contact a second billiard ball, we see the second ball begin to move in a particular direction with a particular velocity. We do not actually see anything that can be called the first billiard ball 'causing' the movement of the second billiard ball. All we observe is the movement of the second ball in a very consistent way whenever it is struck by the first billiard ball in a particular way. If we see this pattern repeated often enough and with sufficiently high consistency, we make the inference that the motion of the first billiard ball causes the motion of the second billiard ball. We have not really seen causality. Rather we have performed an act of inductive reasoning.

Is this way of reasoning valid and justifiable? Hume points out that without direct evidence of causation, the fact that a pattern has consistently occurred in the past provides no philosophical justification for assuming it will continue to occur in the future or even that it becomes more probable as we accumulate past instances of its occurrence. Hume fully understood that in our ordinary life we cannot avoid relying on these kinds of inductive processes to create our expectations about the future. His point is that they are, strictly speaking, not epistemologically justifiable. They are simply psychologically based and involve habits of thought that are inescapable for human beings. This so-called problem of induction has yet to be resolved in a satisfactory way. The twentieth-century philosopher Bertrand Russell has provided a famous example.[17] He considers the case of a chicken who has learned that when he hears the farmer coming each morning, he will be provided with a ration of food for the day. Over time he increasingly comes to expect and anticipate this. One day however, instead of feeding him the farmer grabs him, chops off his head, and serves him for dinner. So much for trusting induction. (For whatever reason, subsequent retellings of this fable typically substitute a turkey for the chicken.)

10.2.2.3 Scientific fundamentalism and universal determinism

Nonetheless, up until the second half of the twentieth century, a law-based view of science formed the background assumption of most of our thinking about what constitutes scientific method. Furthermore, the view that the laws of nature apply everywhere led to the assumption that all behaviour of any sort and in any small corner of the universe, regardless of its complexity, could in principle be explained by appeal to such laws of nature, ideally captured in quantified form as equations. In its boldest form, this view was expressed by the mathematician and scientist Pierre-Simon Laplace who argued that an intellect with infinite capacity and with knowledge of the mass and velocity of every minute component of the universe could predict with precision the precise past and future behaviour of everything. While acknowledging that with our limited human intellectual capacities this could never be fully achieved, he viewed it as achievable in principle.

> We may regard the present state of the universe as the effect of its past and the cause of its future. An intellect which at a certain moment would know all forces that set nature in motion, and all positions of all items of which nature is composed, if this intellect were also vast enough to submit these data to analysis, it would embrace in a single formula the movements of the greatest bodies of the universe and those of the tiniest atom; for such an intellect nothing would be uncertain and the future just like the past would be present before its eyes.[18]

This kind of thinking raised the spectre of full determinism as well as the reduction of all knowledge to physics, which has made science seem such a lethal threat to humanists of every stripe since.

However, given the fact that the portion of the experienced world that we have been able to explain in these ways with any precision is so minute compared to what we have not been able to explain, what is our basis for assuming that this is true? There is currently no scientifically justifiable basis for this assumption. It is a matter of faith, in the words of the philosopher Nancy Cartwright a form of 'scientific fundamentalism'.[19] In this it is reminiscent of the conviction of supporters of EBM that its systematic application will yield consistently better patient outcomes, even though that question itself has yet to be subjected to empirical testing.

10.2.2.4 Are laws of nature literally true?

Furthermore, the purported laws of nature are not literally true of any aspect of the real world.[20] Laws of nature are typically expressed as if the only forces operating are those addressed by the law. This is never literally the case. A plethora of forces and factors are operating in any real-world scenario. It is true that in some situations the magnitude of impact of these other factors are sufficiently small that they can be ignored for practical purposes. It is also true that in some experimental situations, especially in sciences such as physics, the experimental environment can be sufficiently controlled to minimize the impact of extraneous factors. Nevertheless, if one can measure with sufficient precision, the results predicted by the laws of nature are only approximately true. This is often expressed by saying laws of nature require many *ceteris paribus* clauses, that is they are true only with 'all things being equal'.

As an example of the first situation, consider the use of the law of gravitation and laws of motion to predict the movement of the sun and a planet revolving around it. For most purposes applying Newton's law will yield an accurate description of the behaviour of these two bodies. But the law is only applicable in the strict sense if no other forces are at work in the system and the masses of each body are fully concentrated at a point central to their masses. Neither of these conditions are true of the sun and the planets. However, the other forces at work are so much weaker than the gravitational forces, that for practical purposes we can ignore them. But is the law of gravitation meant to be just a useful tool for certain types of calculations of practical import, or is it meant to say something about how the world literally is? If the latter, then the law is in fact false, as the assumptions of the law never really occur in nature. The law is literally true only of a theoretical model in which bodies with point-like masses interact in the absence of any other forces, however small.

In experimental situations, extreme efforts can sometimes be made to shield the interaction under study from other extraneous influences. This can feed the illusion that a law of nature is literally true in the world when it is only an approximation. I recall a discussion with my son-in-law, a quantum optic physicist, in which he was describing his postdoctoral research. His project involved the use of extremely precise lasers and was conducted in a bunker three stories underground to isolate the apparatus from surface vibrations that could create anomalies in the data. Many other devices were in place to exclude the impact of other extraneous factors. Certain laws of physics appeared to apply quite well here, but only because of an extremely artificial and shielded environment from which extraneous forces were excluded. Even here, with sufficiently sensitive measurement, aberrations caused by extraneous influences would keep the laws of physics from being more than very good approximations to real-world behaviour.

10.2.2.5 Unified scientific method and the hierarchy of sciences

A corollary of the worldview that grew out of seventeenth-century science is that while, for practical reasons and due to our limitations, different conceptual tools and methods may be used in different scientific disciplines, there is an intrinsic hierarchy among them. The pinnacle is physics and the belief is that with sufficient knowledge the findings in other fields of science can be reduced to physics. Each discipline meanwhile can be reduced in the direction of physics. Thus, sociology can be reduced to psychology, psychology to biology, biology to chemistry, and chemistry to physics. One upshot of this view of physics as the pinnacle of science is that the methodology of physics, with its heavy emphasis on mathematical quantification and reducing complex phenomena to a small number of laws, is seen as the ideal model for scientific method in all fields. The fact that most fields of science cannot apply this kind of methodology to its subject matter with anything like the yield of physics is usually seen as simply a limitation of our abilities in these areas rather than reflecting the need for a different methodology to match the objects of study.

In a fascinating and thought-provoking book, *The Dappled World: A Study of the Boundaries of Science,* the philosopher of science Nancy Cartwright[21] argues that there is no evidentiary basis for assuming a unified scientific method based on physics. She sees this assumption as another instance of scientific fundamentalism. Rather she makes the case that the subject matter of each area of study requires the development of a methodology appropriate to that subject matter.

10.2.2.6 Logical empiricism and the demarcation problem

Is there a danger here? If scientific methodology is allowed to become so varied, how can we be sure that each of these methodologies is still scientific? How can we distinguish scientific thinking from the full range of the ways people think in their attempt to explain the world, with their varying degrees of validity and rationality? This problem became acutely felt in the early 20th century and was termed the 'demarcation problem'.[22] After all, all sorts of wild theories and explanatory frameworks could be offered about many things. They all could be garbed in language that superficially sounded scientific. How can one distinguish real science from pseudoscience? This problem became more pressing after Einstein's breakthroughs in physics in the early twentieth century. Now science was postulating a view of the world that was radically at variance with common sense. On what basis should it be accepted and, for instance, astrology not?

Philosophers of science like Karl Popper[23] were interested in articulating by what criteria we could judge quantum physics to be science but fields like psychoanalysis and Marxism as not. After all, the latter two also make claims to scientific status. Popper was struck by how seemingly easy it was in the latter fields to explain any contradictory occurrences in terms of the theoretical model. Hence, he developed the view that the essential distinguishing mark of real science was the falsifiability of the hypothesis or theory. No matter how often a hypothesis was confirmed by the evidence, it remained possible that with more evidence it would be contradicted. However, if a hypothesis generated a distinct prediction that with further observation or experimentation was shown to be false, the hypothesis could be rejected with confidence. Popper conceded that at any given time, it might not be practically feasible to test such a prediction derived from a hypothesis. But to be scientific the theory in question needed to be falsifiable in principle. At first glance this appears to be a compelling candidate for demarcating science from pseudosciences. However, on closer examination that line of demarcation becomes a very fuzzy zone for several reasons.

First, Popper's view assumes the ready availability of 'critical experiments' such that their results can only be interpreted one way. In scientific practice, critical experiments are rarely definitive. Many other factors could explain the failure to confirm the hypothesis in any given observation or experiment. There could be equipment failures or errors in measurement. Furthermore, a wide range of other potentially influencing variables that are assumed to be held constant or to be irrelevant to the question at hand may exert an unexpected influence. In short, any hypothesis being tested rests on a bed of many background assumptions that for whatever reason may prove to be inaccurate. One can never test a hypothesis in isolation, but rather one is always testing the full 'web of beliefs' in which that hypothesis is embedded, including all the background assumptions. If the test fails, it does not tell us where in our web of beliefs the error lies. Sometimes it is possible to isolate some individual components with further testing and sometimes it is not.[24]

Second, the notion of a critical experiment is even less applicable in many disciplines, such as medicine and the rest of the biological sciences, where most generalizations that might be viewed as laws of nature are statistical and probabilistic. Thus, if we have evidence supporting the notion that a particular kind of event has an 80% probability of occurring in a particular kind of setting, no one trial can possibly provide a definitive falsification. Indeed, if we apply Popper's standard in psychiatry and medicine, there is almost no piece of therapeutic knowledge for which there is not at least one clinical trial that fails to validate it.

Given that falsification cannot be the basis of distinguishing science from non-science, is there some other way that the scientific enterprise can be characterized that supports its unique validity? The dominant view on this matter in the early and middle twentieth century was developed by philosophers of science characterized as 'logical empiricists'.[25] Many of the leaders of this group were immigrants to the United States from German-speaking nations during and between the world wars. Their position was that we can have confidence in our knowledge only if it is derived from rules of logic or from empirical observation. With the reliance on observation, they represent a continuation of the tradition of David Hume and other British empiricists from the 18th century, blended with an emphasis on formal logic. Furthermore, like Hume, they believe that the fundamental generalizations of science, the laws of nature, were generated by inductive reasoning from multiple observations. Once in place, these generalizations could be viewed as axioms that could be deductively applied to reasoning about and explaining particular instances. From the point of view of the logical empiricists, what identified a theory as truly scientific was the ability to put it into a framework of axioms and logically permissible deductions applied to empirical observations. Note once again, that an individual instance can only be scientifically understood as being a token representative of a categorical type. A hypothesis in this framework would be understood as a posited deduction that could be proven as true or false based on further empirical observations. Relying on these basic principles, several different detailed formulations of this view of science were elaborated, prominent among them were those of Carnap[26] and Hempel.[27]

10.2.3 Problems with the received view and the emergence of model-based science

By the middle of the twentieth century, several serious technical problems became increasingly appreciated in the conceptual framework of logical empiricism.[28] Among others, these included ongoing lack of resolution of the problem of induction and the inability to be able to support a clear categorical distinction between logical and empirical propositions, a fundamental technical issue central to the position of logical empiricism.[29]

Even more telling, especially for students of science who are not professional philosophers, was the growing realization that this view of science as a highly structured axiomatic system did not appear to capture the way scientists themselves thought as they plied their trade. To be fair, the logical empiricists never claimed that scientists consciously proceeded as if applying axiomatic reasoning as they dealt with the day-to-day problems of their scientific specialty. They drew on a distinction first popularized in the English literature by Reichenbach[30] between 'the context of discovery' and 'the context of justification'. The former described the processes by which scientists generated their ideas and specifically their hypotheses. This was seen as essentially a psychological question as to how creative people developed new ideas. The logical empiricists argued that they were interested instead in the epistemological question of how the resulting hypotheses and theories could be theoretically justified. Their position was that for a claim to be justified it must be possible to reformulate the results of scientific work to fit it into an axiomatic system, even if the working scientist was not consciously thinking in those terms.[31]

By the beginning of the second half of the twentieth century, this argument was seeming increasingly strained. A rapidly accumulating body of work in the sociology of science showed, through many case studies of practising scientists, that the path of scientific progress was in part controlled by a range of contingent factors including the vested interests of scientists, their social enculturation into certain schools of thought and the communal interactions of the larger scientific communities to which scientists belonged.[32]

10.2.3.1 The impact of Thomas Kuhn: Paradigms, exemplars, and methodological naturalism

In what has come to be viewed as the start of a seismic shift in thinking about the nature of science, the physicist/philosopher Thomas Kuhn in 1962 published *The Structure of Scientific Revolution*.[33] He argued that looking at the way science is actually practised, even in a field like physics, did not show a gradual steady progression in scientific knowledge that resembled elaborating and filling in the details of a coherent axiomatic system that progressively approximated the truth. He argued instead for an evolution of science marked by a series of stages. A period in which radically different frameworks for understanding a discipline openly competed was followed by a period in which one framework became the dominant paradigm. This brought in a period of 'normal science' in which scientists worked on solving specific problems or puzzles to fill in more and more of the details of the dominant paradigm.

The paradigm, in its turn, suggested certain very typical types of problems with typical types of solution, which Kuhn called exemplars. During periods of normal science, a scientist confronted with a new problem would look to see if it resembled any of the exemplars that had come to typify that paradigm. The scientists would attempt to use the strategies embedded in those exemplars to solve a new problem. The training of scientists involved familiarity with a range of exemplars in the field and experience in applying these to new problems. For a productive paradigm this could lead to a long period of constructive additions to a science.

Inevitably however certain problems prove stubbornly resistant to the strategies provided by the exemplars. Credible scientific experiments replicate findings inconsistent with the prevailing paradigm. Some of those problems suggest important internal inconsistencies in the dominant paradigm, which Kuhn termed 'anomalies'. As anomalies mount, increasing doubts about the dominant paradigm grow. At some tipping point the anomalous findings lead a subset of scientists to suggest an alternative paradigm resulting in a period that Kuhn termed 'scientific revolution'. Often the new contending paradigm is supported by younger members of the scientific profession, while older members tend to remain loyal to the older paradigm in which they were trained and in which they practised. If the new paradigm deals more adequately with the anomalies, it might in time become the new dominant paradigm with a return to normal science.

Kuhn crucially argued that there was no atheoretical position from which to judge one paradigm as more correct or truer than another. The paradigm adopted by scientists places imaginative constraints on how they view and interpret the evidence, even including their judgements as to what counts as relevant evidence. Each paradigm is likely to be better at solving certain types of problems, and what paradigm appears superior at a given moment depends heavily upon which problems practising scientists are most interested in at that time. From Kuhn's perspective this evolution of science bears no resemblance to the process of an unfolding axiomatic system as proposed by the logical empiricists.

The work of Thomas Kuhn was highly ambitious and represented a radical departure from previous ways of thinking about science. It has spawned a huge literature and a host of discussions about various aspects of the nature of science. For our purposes however, we can focus on just a few implications. First, the position that science can justify what it does by appealing to how its observations fit into an axiomatic system no longer seemed possible. The logical empiricists viewed all non-analytic knowledge as derived from experience. One first makes observations and then develops inductive generalizations from which an axiomatic system is generated. The system then becomes the basis of explaining subsequent observations. To work as intended, this approach requires that observations be uncontaminated by theoretical assumptions. What can count as empirical evidence for the theory must be independent of the theory. But Kuhn's notion of paradigms argues that the paradigm itself heavily determines what counts as legitimate evidence and how that evidence is collected and characterized. He went so far as to argue that since different paradigms see evidence differently, no observations could be used to definitively argue for one paradigm over another. While this more extreme position has been criticized, the fact that all observation is theory laden to varying degrees is generally accepted and is enough to undercut the earlier understanding of science.

Second, if an accurate understanding of scientific method can no longer come from an armchair theoretical discussion of logical relationships, a meaningful philosophical understanding of science must look carefully at what practising scientists actually do and how they think about that process. This approach is itself empirical in spirit, and often is called 'naturalism'.[34] From the perspective of naturalism, the goal of a philosophical understanding of science is not to demonstrate an airtight logical justification. Rather, naturalism's goal is to understand and characterize scientists' practices and procedures to account for the breathtaking successes science and technology have enjoyed from the seventeenth century to the present. To explain the success of science in naturalistic terms does not require that scientific reasoning can be shown to be categorically distinct from other valid forms of human reasoning. Indeed, naturalistic philosophers of science frequently draw on work in cognitive psychology to show the commonality of scientific reasoning with processes of effective human reasoning in general.[35] The question shifts from what makes science unlike and distinct from other kinds of human reasoning. Instead, the issue is to identify the rigours and constraints that have allowed the instances of human reasoning that we call science to be so remarkably productive.

Third, exemplars as understood by Kuhn indeed appear to play a pivotal role in the way science is practised across scientific disciplines. Studies of how scientists actually proceed in solving problems demonstrate Kuhn's position that, rather than relying on axioms or first principles, practising scientists typically look to see if the problem they are currently addressing resembles any of the model problems and solutions they were taught to use in their training in their specific scientific discipline. The scientist then attempts to use such models as a guide to developing potential solutions to the current problem they are seeking to solve.[36]

If the goal of science is not the construction of an axiomatic system, how else can the purpose of science be conceptualized? Most scientific activity appears to be guided by two types of goals: explaining or representing on the one hand, and predicting or intervening on the other.[37] The meaning of these terms will be fleshed out in the discussion that follows.

10.2.3.2 Science as practised: Models and families of models

The view of science that I will put forth here is based on this naturalistic perspective. I am drawn to this view both because it seems to me to accurately characterize and explain how science is practised over a wide range of scientific disciplines, and because it is especially effective in understanding and justifying methods in applied sciences such as clinical medicine.

Ronald Giere argues that an effective place to look at how scientists reason in their day-to-day work is to examine how a particular scientific discipline is taught to its advanced students. Hence, he has looked at advanced textbooks. Having trained in physics before entering philosophy, his favourite examples are textbooks of classical mechanics. He points out that if one looks at the most widely used textbooks of classical mechanics for advanced students, they generally start with the statement of Newton's laws of motion.

> Now an important fact emerges. From the second law, or from all three, nothing much of interest follows. If the laws of motion by themselves are taken as the axioms of an axiomatic system, it is a rather uninteresting one. Nor do any texts I know proceed to deduce theorems from these 'axioms'. Rather, they typically proceed to a series of chapters based on different assumptions about the form of the force function . . . It is force functions that provide the chief organizing principle in most classical mechanics textbooks . . . Whatever the particular order of presentation or groupings of topics, there is a remarkably common set of core topics and examples. The conclusion that many of these core examples are the historical descendants of Kuhnian exemplars is irresistible.[38]

Different examples with their associated force functions describe different theoretical models, with varying degrees of abstractness, that can be applied to certain real-life situations. A group of similar models at one level of abstraction can in turn be described by a still more general and more abstract model. In this view a scientific theory is not a set of axiomatic statements or laws of nature. Rather they are clusters or families of models, or families of families of models, that have shown themselves to be useful in explaining various concrete situations in the real world. In the case of classical mechanics for instance, the most abstract family might represent the conservative versus nonconservative models.[39] Each in turn contain less abstract and more specific model families such as models of rectilinear motion, models of harmonic motion, and models of orbital motion. More concrete models are contained in each of these categories, such as freefall and inclined planes for rectilinear motion, pendulums and springs for harmonic motion, and circular orbits versus elliptical ones for orbital motion.[40] Finally at the most concrete level are models meant to represent specific and individual instances, for instance a model of a particular real-world device characterized as a pendulum or an inclined plane or as a combination of such elements.

To make this notion of families of models with varying degrees of abstraction more relevant to psychiatry, one could speculate about moving beyond the kind of POP models of individual patients we have been discussing, towards developing models that attempt to capture the salient variables in understanding a particular diagnosis. One might develop a model for major depression, for bipolar disorder, for generalized anxiety disorder, for obsessive compulsive disorder (OCD), for psychopathy, for borderline personality, etc. Obviously, we are not yet at a point where well-articulated and useful models exist in a way comparable to what we find in physics. But one can imagine what such models would look like. One then could imagine more abstracted and generalized models, such as a model for affective disorders or for psychotic disorders. Such models would presumably capture still more abstract descriptions. We may see patterns of similarity that would allow even more abstract models to be developed, for instance models of internalizing disorders versus models of externalizing disorders. None of these models would be literally true of any specific individual. Their value would come from the extent to which they presented possible structures of interacting propensities, some of which could be used by the clinician in constructing a very concrete model for an individual patient. Very preliminary and crude versions of these kinds of more abstract models can be constructed based upon what our current understanding of these conditions is, even if they are not yet usually characterized in terms of models.

Another example that is more fully established, and has considerable relevance in psychiatry, can be seen in our understanding of neurotransmitters. As various neurotransmitters have been discovered and characterized, models for how they are manufactured and stored in cells, mechanisms that control their release and reuptake, how they interact with receptors and how they bring about downstream effects have been developed. While there are some differences in all these features between different neurotransmitters, there is enough commonality that one can construct a more abstract model as to how neurotransmitter systems work in general. That more abstract model does not literally apply to any specific neurotransmitter. Its value lies in the fact that, upon the discovery of new neurotransmitter, a scientist can look to see the extent to which those more abstract patterns might provide a guide to understanding how the novel neurotransmitter operates and help formulate a hypothesized model for that transmitter's behaviour. It would be optimally useful if we could reach a point where we could use a mid-level model of a particular neurotransmitter to help develop a concrete model of how that neurotransmitter operated in a specific individual.

Based on this naturalistic examination of how scientists are educated in their field and trained to apply that education, Giere proposes the following:

> We understand a theory as comprising two elements: (1) a population of models, and (2) various hypotheses linking those models with systems in the real world. Thus, what one finds in textbooks is not literally the theory itself, but statements defining the models that are part of the theory. One also finds formulations of some of the hypotheses that are also part of the theory.[41]

10.2.3.3 The relationship of models to reality

However, in what sense does a hypothesis link models with systems in the real world? Even the most concrete model intended to describe a specific state of affairs in the real world is never literally true of any aspect of the real world. At best it is an approximation. Many aspects of the real-world system of interest are ignored as irrelevant or of insufficient magnitude for the problem at hand.

> Hypotheses, then, claim a similarity between models and real systems. But since anything is similar to anything else in some respects and to some degree, claims of similarity are vacuous without at least an implicit specification of relevant respects and degrees. The general form of a theoretical hypothesis is thus: Such and such identifiable real system is similar to a designated model in indicated respects and degrees . . . To claim a hypothesis is true is to claim no more or less than that an indicated type and degree of similarity exists between a model and a real system. [As an example] . . . the positions and velocities of the earth and moon in the Earth–Moon system are very close to those of a two-particle Newtonian model with an inverse square central force.[42]

How does one decide what level of similarity, in what respects and what degrees, should apply in judging a specific hypothesis? No standard or objectively fixed answer to this question exists. The nature and degree of the required similarity between a model and a system in the real world depends upon the task at hand. This necessity for specifying a task at hand makes model-building a practical endeavour.

Thus, a model is a kind of map. How precise the map needs to be and in what ways it needs to resemble the piece of the world it seeks to represent depends totally on the purpose for which the map is being used. 'It is not the model that is doing the representing; it is a scientist using the model who is doing the representing.'[43] Furthermore, the scientist decides the model's fitness in serving one's purposes. Notably, if the real-world system one is attempting to characterize is complex, one is likely to need to use multiple abstract models and combine elements of those models to generate a more specific and complex model of that concrete real-world system.

> In actual cases, multiple causes are likely to be present and interact, and other local elements may also contribute to a specific causal history. Thus in an explanation, models of various possible contributing factors must be integrated to yield the correct description of the actual constellation of causes and conditions that brought about the event to be explained.[44]

10.2.3.4 Studies of scientists at work

Beyond the structure of advanced textbooks, the central role of models in the way scientists actually work is also supported by studies of how scientists go about solving a problem.

> One of the main devices employed in those studies is to have a scientist think aloud while solving a problem. The researcher then analyses the protocols generated by the scientist. From the few published reports it seems that physicists do not begin by writing down Newton's equations, adding a force function, and deducing. Rather, they first select from memory a representation of the problem—a model—and work from there. What is retrieved from the physicists' long-term memory, it seems, is not the axioms of mechanics, but an appropriate model.[45]

Furthermore, this way of thinking becomes more prominent as a scientist's experience increases. Note the striking similarity between this characterization and the research described in sections 8.13–14, that examines the development of expertise in decision-making in general by employing pattern recognition and simulation.

This characterization of scientific theorizing as centred on the use of models to characterize selected aspects of the real world applies not just to physics, but to any scientific discipline.

> The main general lesson is: When approaching a theory, look first for the models and then for the hypotheses employing those models. Don't look for general principles, axioms, or the like. Of course, pay close attention to what scientists write and say. Models are typically characterized using words, though diagrams are often equally important. But language is merely a means to an end—the characterization of a population of models[46]

What distinguishes physics is the high level of mathematical quantification of most models. This indeed adds considerable power to the predictive force of the models, but is not essential to how models function. In many other disciplines, models are much more qualitative and approximate, but nonetheless are central to whatever degree of predictive power those disciplines possess. While many philosophers of science focus heavily on physics for their examples, this general framework applies across disciplines. For example, Giere has also looked at geology,[47] Cartwright at economics,[48] and Mitchell at biology.[49]

10.2.3.5 Models of particulars ground confidence in science

However, if the validity of a science cannot be judged by its capacity to establish an axiomatic system of general laws of nature, how do we judge the validity of a discipline construed as a family of models? Another way of asking this question is what provides the rigour in this view of scientific method that explains the extraordinary success of science? This turns out to be a layered process. The most fundamental level of validation is grounded in the most specific and concrete models that are constructed to characterize unique real-world systems. Such a specific model posits a number of factors and their interaction. We previously argued that these factors are best thought of as propensities, a topic that we will return to shortly. Such a model can be applied in a backwards direction to explain the current state of that real-world system—how it got to be the way it is now from the interaction of the relevant propensities in the past. This is science as providing explanation and representation. However, by looking forward, the same propensities and their interaction can yield predictions about the future behaviour of that real-world system, including the response of that system to new external perturbations and interventions. Using the model to generate predictions provides for an assessment of its validity. This kind of validity has a strong pragmatic bent. Remember that models are not literally true or false with respect to the real world. Rather the question is to what extent they successfully resemble the real-world system and if they meet the practical intention of the scientist. Those questions can be answered only with respect to the goals of the individual utilizing the model. The question becomes: does this model when applied to this real-world system for this specific purpose bring about the desired result with sufficient accuracy for the task at hand?

Note that the epistemological challenge has been stood on its head in contrast to the older view of scientific method and its justification. The fundamental prediction is now at the level of the individual instance and the adequacy of the model is assessed through that individual instance. As a discipline evolves, certain commonalities among models yield more generalized, encompassing, and abstract models that constitute families of models. This process can be repeated to yield families of more generalized models still. Certain principles may emerge that appear to be true of large categories of models. These become the generalizations of the discipline in question.

How do we then judge the validity of these generalizations and of the more abstract models? These are even more removed from concrete reality than the specific models. They are not literally true of the real-world, but are only true of the less abstract models that use them. It follows that the broader generalizations and more abstract models can only be judged by their fruitfulness in generating successful specific models to address new concrete real-world problems. The rubber meets the road only in the context of a prediction about a concrete, real-world situation that comes out of the use of an individualized model of that situation.

10.2.3.6 A science of the individual is possible

Most of what scientists spend their time doing in any discipline is trying to solve concrete and specific problems. In this framework then, a science of the individual case is not only a possibility, but indeed is the fundamental level upon which the justification of any scientific discipline can be built. For instance, if we had a good model for the development of major depression, how could that general model be epistemologically justified? It would need to generate numerous concrete models of specifically depressed individuals that sufficiently resembled the particular real-world situations of those individuals to prove adequate in predicting their behaviour and response to interventions with sufficient accuracy for the purposes at hand. This parallels the case in physics where Newton's general laws of motion are epistemologically justified because they provide useful blueprints for developing more specific models of concrete situations across a wide range of families of models, including instances of rectilinear motion, harmonic motion, and orbital motion. Likewise, the principle of gravity is justified because it facilitates the construction of models that predict, with sufficient accuracy for many purposes, the behaviour of our specific Earth–Moon system, our specific solar system, and numerous concrete interactions between specific objects on earth and in the cosmos.

10.2.3.7 The assessment of competing models

In practice, when one is looking to validate a specific model of a specific real-world situation, it is usually in terms of how that model compares with some alternative model. It is only when a viable alternative model exists that the question of epistemological justification is of more than theoretical interest. The question typically is: Why should we accept model A for explaining and making predictions about real-world system X as opposed to model B? If both models are in fact viable contenders, they should generate plausible explanations for how the current state of X came to be.

A realistic way to judge one model epistemologically superior to the other is if it generates differing predictions about the future behaviour of system X, or of the response of system X to some intervention. The structured application of such an intervention and the careful assessment of its consequences is, of course, what we call an experiment in science. If we wish to generate an epistemological justification for positing a very specific causal factor, we need the experiment to be conducted in a way that controls for as many other variables as possible, and shields the experimental situation from unspecified extraneous influences from the larger environment. This is much more achievable in some scientific disciplines than in others. Fortunately, such experimental purity is not always necessary if our goal is deciding whether model A or model B best represents system X for our purposes. If the predictions generated by model A sufficiently resemble what happens in system X when model B does not, then model A is justified as our best current working model for predicting the behaviour of X. Of course, if A in the future fails in predicting other aspects of interest in the behaviour of X, A must be revised or an alternative model generated.

In practice, there may be reasons other than epistemological ones for choosing one model over another. Simplicity, ease-of-use, and cost may be relevant. Consider the treatment for erectile dysfunction (ED). Prior to the development of sildenafil (Viagra) and other phosphodiesterase 5 (PDE-5) inhibitors, treatment of ED first required making a diagnostic distinction between organic and psychogenic causes. For the latter, the time-consuming process of psychotherapy was the treatment of choice. With the appearance of PDE-5 inhibitors, the treatment of ED became simpler and more efficient. The distinction between organic and psychogenic origins of ED became less important in many cases, and the response to the medications was more rapid than to psychotherapy. Thus, the use of PDE-5 inhibitors became the first-line treatment of ED not primarily because it was thought to represent a more accurate explanatory model, but because of its practical advantages.[50]

10.2.3.8 The nature of propensities

We will return shortly to a consideration of how this naturalistic approach to understanding scientific activity applies to the question of the epistemological justification of POP models in psychiatry. Before doing that, however, I want to discuss one other rationale for our choice of propensities as the way we have characterized the components of models. I argued for the term 'propensities' as an alternative to 'causes' to describe the interacting elements of our POP models. I based that on the intent to remain neutral as to how strictly deterministic the interactions between elements are and what room should be left for human choice and existential freedom.

A further reason makes propensities a good choice. It turns out that propensities are a useful conceptualization for model components not just in fields like psychiatry where human action is in question, but even in models in fields such as physics. A propensity-based understanding appeared in the later work of Karl Popper.[51] More recently this approach has been elaborated and developed by Nancy Cartwright.[52] Since David Hume, the notion of causation has been heavily tied to the idea of one event (E), the effect, consistently following from another event (C), the cause. However, in many complex real-world situations, C occurs but is not followed by E. This often can be understood because other factors of influence are present that somehow block the action of C. This is easiest to appreciate when the blocking factors are well known or obvious. For instance, while gravity is certainly exerting influence on a flying airplane, other well-described factors block the airplane from falling to the earth, which is certainly an effect that would be predicted if gravity were operating in isolation. In many other cases, however, the other factors that interfere with C resulting in E are unknown. This is very common in medicine. Thus, we say that smoking causes lung cancer, yet most smokers never develop lung cancer. We sometimes say in cases like this that the relationship between cause and effect is statistical or probabilistic. But this is a quite different situation from saying that the outcome of a roll of the dice is statistical or probabilistic. I would suggest that what we really mean here is that in any specific case there are multiple other variables that either synergize with smoking to increase the likelihood of cancer or work against the effect of smoking such that cancer does not develop.

In even minimally complex systems, causation cannot be understood as a law-like regularity of E following C. It is more helpful to think of causation as a propensity—a tendency, disposition, or capacity inherent in the element in question—to bring about a certain result. Whether this propensity leads to that result in any specific situation depends upon the other elements and their propensities as they operate in the system. Thus, gravity is the propensity or disposition for bodies with mass to move towards one another with a specific rate of acceleration. Whether that occurs in any real situation depends upon what other propensities are operative. So, if the forward motion of the airplane allows the forces of air resistance to operate, the propensity of the airplane to fall to the ground is not actualized. If the engines fail however, the propensity of the plane and the earth to rapidly move towards one another is tragically realized. Propensities then are attributes of structures in the real world that are inherent in the nature of those structures, whether they are actualized in any particular situation or not.

> Laws in the sense of claims about what regularly happens are not our most basic kinds of scientific knowledge. More basic is knowledge about capacities, in particular about what capacities are associated with what features. (Or, put more laboriously, 'knowledge about what capacities the system will have in virtue of having particular features'.) . . . Knowledge of capacities is more basic in that it is both more embracing and more widely useful than knowledge of regularities.[53]

Cartwright makes the interesting observation that this view of causality was well captured by Aristotle in his concept of natures. Because early modern science developed in and as a rejection of a world dominated by rigidly scholastic theories based on Aristotle, a wholesale rejection of Aristotle's science occurred[54]. A number of quite valuable babies went out with that bathwater: ' . . . we get a far better description of scientific knowledge if we adopt a category of Aristotle's. The knowledge we have of the capacity of a feature is not knowledge of what things with that feature do but rather knowledge of the nature of that feature'.[55] Cartwright goes on to note that in this borrowing from the Aristotelian idea, the concept of a 'nature' is changed in several important ways from his understanding. Unlike in Aristotle's view, natures are assigned

> . . . not to substances but rather to collections or configurations of properties, or to structures. . . and that what appears on the surface is a result of the complex interaction of natures . . . [H]aving made the empiricist turn, we no longer identify natures with essences . . . Still, when we associate a particular principle of change with a given structure or characteristic, we expect that association to be

> permanent, to last so long as the structure is what it is. Indeed, it is this permanence of association that I underline by claiming that modern science still studies Aristotelian style natures. Of course these are not really Aristotelian natures. For one thing we seem to share none of the concerns about substance and individuation in which Aristotle's concept was embedded.[56]

10.2.3.9 Rethinking our view of science

This shift to a model-based view of science generates a fundamental rethinking of what constitutes scientific method, and what makes it rigorous, successful, and deserving of our epistemic confidence in its results. That confidence no longer derives from establishing encompassing generalizations in the form of laws of nature. Instead, it derives from its success in generating accurate predictions, especially predictions as to how the concrete system under study will change in response to a particular intervention. The critical ground level of this process comes from predictions about the specific real-world systems under study. Patterns of predictions across multiple such systems, if consistent, can support our confidence in more generalized models applying to categories of concrete systems. In psychiatry, the concrete systems are the individual patients and the problems they bring to us for treatment. Broader categories could be framed in terms of diagnoses or other variables that characterize a group of individuals.

10.2.3.10 An interventionist view of causation and explanation

From a naturalistic perspective, the core goals of science are explanation and prediction. We have discussed the central role of prediction in model-based views of science. It also requires a fundamental shift of how we think about explanation. At its most basic, explaining a phenomenon is about saying what caused it to come about. In the view of science that evolved from the time of Newton through the first half of the twentieth century, the answer to that question took the form of showing that the phenomenon of interest was an instance of a more or less generalized regularity or law of nature.

In rejecting that formulation, what do we now mean in saying that A causes B? While at first glance this may seem like a self-evident and non-controversial question, it turns out to be quite challenging and lends itself to multiple frameworks of understanding. Hume made the point that causality itself cannot be observed. It must be inferred based on other observations. One approach that has received considerable recent attention has been termed a 'manipulationist' or 'interventionist' account of explanation and causation. A particularly comprehensive discussion of these issues is presented by the philosopher James Woodward.[57] By this account, to say that A is a cause of B is to assert that some possible or hypothetical change in A would result in some change in B. Phrased differently, to be a valid explanation, it should be possible to use it to answer a 'what-if-things-had-been-different-question'.

Clearly this view of explanation is simply a backward-looking version of prediction. In this light, consider an experiment involving an intervention using factor I to alter the state of factor A and making a prediction about changes in the future state of B. If B changes in response to I's alteration of A, we are not only assessing the utility of intervention I. We are also asserting that factor A is part of the causal complex responsible for B. For example, in the case of Ms M, her work with the female therapist on her distorted sense of self deriving from her relationship with her parents resulted in a marked change in her capacity to sustain a relationship with a man. Viewed as an experiment, this not only argues for the efficacy of the intervention, but also argues that those parental relationships had some causal role in bringing about her difficulties. Similarly, Ms K's reduction in anxiety in response to a serotonergic antidepressant not only argues for the efficacy of that drug in her case, but also argues that something about the state of her serotonergic system played some causal role in her anxiety. It is important to note that in neither case are we arguing that the propensity influenced by the intervention is the sole cause, but simply that it is part of the causal or explanatory matrix.

This approach to causation allows one to argue that A is a cause of B even if we do not fully understand the mechanism by which that factor's influence is brought about. Working out the full conceptual details of this approach to causation is a formidable task that goes beyond what we can discuss here. There are however several points about the interventionist approach to causal explanation that are worth mentioning. The pragmatic quality of this approach makes it very appealing in understanding causation in applied as well as basic sciences. It also fits very readily with thinking about causation in an individual, without relying on large-scale regularities covering groups of instances. It applies to a broad notion of causality that includes both deterministic factors and exercises of free choice.

This lack of a commitment to a specific kind of mechanism also makes it a useful approach in fields, like psychiatry, where our understanding of underlying mechanisms is often very sparse. More specifically, it fits extremely well into the way of thinking that underlies POP model construction, with its emphasis on patterns of interacting propensities that are posited to impact the state of a specific individual. The potential value of an interventionist approach in psychiatry has been compellingly put forth by Kendler and Campbell.[58]

10.3 Back to psychiatry

After this detour into the history and current framework for understanding scientific method, it is time to first summarize where we have come and then refocus on how this applies to our primary area of interest, the POP methodology of clinical reasoning and decision-making in psychiatry proper. In this chapter we are specifically interested in addressing the epistemological question of how justified we are in accepting the conclusions drawn from these models of our individual patients as valid knowledge. It was suggested that in our modern world where science is accorded high epistemological status, and this is certainly true in contemporary medicine and psychiatry, it becomes an important question whether the generation of a model of an individual patient is a scientific activity or something that stands apart from science.

As a preliminary to addressing this question we have looked at two broad frameworks for characterizing what makes an activity scientific. The first, deriving from logical empiricism and often termed the 'received view', dominated thinking about science in the early and middle years of the twentieth century. It had so dominated that it often remains the implicit and assumed notion of what it is to be scientific both among some scientists and the public at large. I have argued that it represents the implicit background assumptions behind EBM. But starting in the second half of the twentieth century, problems with the received view have been exposed. Alternative approaches to understanding science have subsequently appeared. I have described one widely shared framework, in which models play a central role, that has become an important position in the contemporary understanding of science. I have tried to make the case for why I find this view compelling. Nonetheless, there certainly are competing positions in the field. It is not necessary to definitively demonstrate the correctness of the latter position over the traditional view to argue that the traditional standards should not have the unquestioned authority to define what is and what is not scientific.

10.3.1 The science of psychiatry

The traditional view of science as an inductive process that generates abstract generalizations or laws which in turn are applied deductively to the individual case would argue that unique modelling of an individual patient cannot be scientific. We have argued for a radical revision of our view of scientific method in which the generation of particular models to represent concrete pieces of reality is in fact the central scientific activity. To the extent that increasingly abstract and generalizable models can be generated from the collective experience with a range of individual situations, this is desirable and useful. Its use however is not primarily in discovering deep universal truths about nature, but in providing useful tools to more effectively generate new models to address more and more concrete situations. The central task of science, and the primary measure of its value, lies in the ability to solve concrete problems.

This is precisely what the psychiatrist does in generating a POP model of the individual patient who is seeking help. As such, I argue that what the psychiatrist does in working with the individual patient in this way is not simply compatible with scientific method, but represents a clear application of it. The fact that the psychiatrist's immediate domain of interest is limited to a particular patient does not make this activity unscientific. When one is treating Ms Jones, one is truly engaged in the science of 'Jonesology'. The model being generated is meant to represent relevant aspects of the distress that has brought Ms Jones to treatment. If it is a successful model, it should provide a plausible explanation of how Ms Jones' distress came about, and more importantly, generate concrete predictions of how that distress will evolve under the influence of possible interventions.

The psychiatrist will draw on resources of two types to assist in constructing a workable model out of the mass of unique information acquired from a careful history and examination of the individual patient. To the extent that Ms Jones' situation resembles other patients, either previously seen by the psychiatrist or learned about from other psychiatrists during training or from reading the literature, elements from models used in those cases may suggest themselves as being applicable to Ms Jones. Some of these models have enough utility in generating models for new patients that they can be regarded as exemplars in Kuhn's sense of the word. Secondly, the psychiatrist can draw upon more abstract generalized information about groups of patients or from studies of basic mechanisms that are relevant to Ms Jones. In this case the psychiatrist is drawing upon knowledge not from psychiatry proper, but from other closely related informing disciplines such as clinical therapeutics, psychopathology, psychology, neuroscience, biology, sociology, and anthropology. This sort of information can suggest certain propensities and their interactions that might be usefully incorporated into a unique model of Ms Jones.

A diagnostic category, for instance, if it is to have value to the practising psychiatrist, essentially serves as a mid-level generalization that can suggest certain features and interactions that can be incorporated into a unique model. From this perspective the many debates about whether psychiatric diagnoses are real entities in the world, or are natural kinds, or cut nature at its joints, may in fact be relatively unimportant. They should be replaced by the question as to whether they are useful as generalizations in generating more effective models of individual patients. As we have seen from some of the examples discussed earlier, diagnoses in fact often play a useful role in contributing elements for concrete models of patients when more specific understandings of the underlying mechanisms have not been well characterized but are thought to exist.

One of the conclusions of the contemporary naturalistic philosophy of science is that we have no reason to expect a single uniform methodology or approach to model construction for all the varied scientific disciplines. Rather the subject matter of each discipline will help to define the most appropriate methodology. Each discipline will develop a process of model construction that best fits that subject matter. I would argue for the conclusion, based on all that is gone before in this discussion, that POP modelling is the appropriate methodology for psychiatry proper. That methodology is distinct from the appropriate methodologies for the informing disciplines that surround psychiatry. Each of those must find its own best methods that suit the problems in each of those fields. Several preliminary efforts to apply methodologies of model construction to circumscribed areas of psychiatric practice, such as reaching a diagnosis[59] and conducting psychotherapy,[60] have appeared in the recent literature.

10.3.2 Establishing confidence in our models by testing predictions

Granted that what the psychiatrist does in generating a POP model for an individual patient is an instance of the fundamental activity of any scientific field, we still must address the question of how confident we can be in accepting any given model of a patient as valid. We have seen that the epistemological strength and credibility of science comes not from an axiomatic system of general and universal laws of nature, but rather from the ability of specific models to generate sufficiently accurate predictions about the behaviour of a concrete piece of the real world to meet the needs of the task at hand. Hence if the model-generating activity of the practising psychiatrist is to have credibility, one must look carefully at how successful predictions based on that model are. I believe that one of the frequent shortcomings that makes psychiatry often appear impressionistic and lacking in rigour is that we are often reluctant to generate specific predictions in a sufficiently short time frame to be practically useful in assessing our models. If the multiple clinicians treating Ms W (section 9.7) had viewed their failed treatment trials as predictions from their model for understanding her, they would likely have come to question their model much sooner and spared her an extended period of clinical deterioration.

10.3.2.1 Impediments to prediction in psychiatry

Many factors contribute to this failure. In some treatment modalities, especially older psychotherapeutic approaches such as psychodynamic psychotherapy, clinicians have been trained to think in long-term global terms rather than generating specific short-term predictions. I argued that even when goals are reasonably global and long-term, it should be possible to generate specific short-term predictions as to how the patient will behave or respond if the model is correct. Psychiatrists need to be trained to do this in a proactive way. The more predictions generated and tested the greater the epistemological credibility of the model. This can be done on a very fine-grained level. What do we anticipate if we choose to make this specific interpretation at this specific time to the patient? If this response fails to occur, what conclusions should we draw and should we modify or reject our working model? Thus, in the case of Ms M (Chapter 6), a particular type of transference in her initial relationship to me as her therapist was predicted based on the model. That proved to be true. I also predicted that she could sustain the therapeutic alliance despite that transference and stay in treatment. That prediction proved to be inaccurate, and resulted in a change in the model leading to suggesting that she work with a female therapist. Even more concretely, in the case of Ms V, the model was used to generate a prediction about a very specific interpretation that proved to be effective.

Another factor that discourages rigorous prediction in psychiatry is the increasing fragmentation of psychiatric treatment. A resident or hospitalist working on a busy inpatient service often gets very little feedback as to how patients' stays on the unit influence even their short-term outcomes. System changes to allow more ongoing care for patients, and where that is not possible, systematic feedback as to patients' short-term outcomes would be extremely helpful. This could take the form, for instance, of clinical case conferences that included both inpatient and outpatient clinicians treating the same patients. As an additional benefit, any format that allows multiple clinicians to assess a model for a given patient brings to bear the expertise and increased objectivity of the community of scientists, thereby adding to the rigour of the process.[61]

Finally, as discussed earlier, the process of assessing whether a model should be seen as credible in practice usually involves comparing it to an alternative viable model. Hence a useful exercise for the psychiatrist would be to ask a hypothetical question: If the model I have of Ms. Jones were wrong, what do I think would be the next best plausible model and what would be the differences in predictions based on that model compared to the one I am championing?

10.3.2.2 Interpreting failed predictions and POP model revision

Any failed prediction can be interpreted in multiple ways. Sometimes a modest adjustment to the model is in order. At other times the model should be jettisoned and an alternative developed in its place. Often the best course of action is unclear. However, this kind of uncertainty is true in all fields of science. As we discussed in section 10.2.2.6 in looking at the views of Karl Popper, there are in practice very few truly definitive experiments that invalidate a model once and for all. The best one can hope for in psychiatry or any other science is to subject revised models to further predictions and revisions.

Iterative processes such as this one can lead us to progressively more powerful and useful understandings of the world, understandings that are the foundation for the remarkable achievements of the many fields of modern science. The challenge for the busy psychiatrist is to continue to think of every model of every patient as an evolving scientific theory about that corner of the world that is Ms Jones or Mr Smith, and more specifically about those aspects of patients' experiences and behaviours that are relevant to the distress that brought them to treatment. Every prediction, in turn, should be viewed as an experiment that tests the adequacy of the theory.

10.3.3 The practising clinician as scientist

There is another important implication for model construction. The act of constructing a model of the individual patient by a treating psychiatrist is an instance of the quintessential activity of the scientist. We have seen that the construction of unique models of particular problems confronting the working scientist forms the foundation upon which more abstract and generalized scientific models are built. Furthermore, the justification for these more general models lies in their ability to be useful in iteratively generating more productive models for more concrete problem-solving in new unique cases. As a result, fruitful development in any scientific field requires avenues of communication among the community of scientists to be informed about and to learn from the host of concrete models generated so as to suggest hypotheses about more abstract models and generalizations.

Traditionally in psychiatry this was done through a rich tradition of published case studies and case series where individual features of a patient or a small group of patients could be characterized in some detail. Knowledge development could then be a recursive process between concrete models generated by clinicians treating individual patients and more selected formal studies of generalized questions that could provide relevant and useful information back to the clinician.

10.3.4 Reestablishing the circle of knowledge between clinician and researcher

With the rise of EBM, the evidentiary status of knowledge derived from detailed presentations of individual patients has been seriously degraded, reflected in the near disappearance of case studies from the mainstream psychiatric literature.[62] As a result, the propagation of knowledge in the field tends to be seen as a one-way street. Professional researchers conduct studies on groups of patients usually resulting in some sorts of statistical generalizations. These in turn are seen as knowledge fed to the practising clinician who should apply it to the individual patient one is treating. With the absence of knowledge flowing in the opposite direction, from the concrete clinical experience back to the more abstracted work of the researcher conducting more formalized clinical trials, it is not surprising that all too often the research literature appears to have so little meaningful application to the complex individuals that the clinician is trying to help.[63] Nor is it surprising that the actual advances in the treatment of our patients in recent decades has consistently fallen so short of expectations. The circle of scientific knowledge between the more particular and the more generalized desperately needs to be reestablished in psychiatry.

One potentially very useful tool in helping to close this circle in scientific knowledge between the particular and the general has been around for a long time but curiously has gained remarkably little traction, the N of 1, or intensive single-case study. One can look upon this approach as an attempt to impose a higher level of rigour on the process we have described in applying POP modelling. We discussed the importance of making discrete and timely predictions to test our working model for the patient. If our prediction proves to be accurate, we at least tentatively conclude that our model is adequate. However, we could go wrong several ways. If only one instance of applying an intervention yields positive results, this could be a coincidental occurrence having nothing to do with the intervention itself. But we have noted that very often patients present with an ongoing or recurrent problem for which the intervention must be made in an ongoing manner or intermittently reapplied. If it continues to prove effective, our confidence in our understanding of the patient and our interventions correspondingly improves. But other factors could explain the apparent effectiveness. The patient's expectations may lead to a placebo response. The psychiatrist's expectations may lead to a biased assessment of the patient's progress.

In assessing a placebo response, though, the question has actually changed. The treatment has clearly worked, but we are left with the question of why it worked. Did it work because of specific properties of the intervention or nonspecific changes induced by the patient's expectations? In many practical real-world situations this may in fact not matter much. But in circumstances where treatment will be ongoing and carries a significant potential burden in terms of adverse effects or cost, however, it matters a great deal. It also matters a great deal if we want to offer the experience with this individual patient as contributing useful information that might have relevance to other patients and their treatment or as a guide to future research questions.

This is where the N of 1 study can play a valuable role. In this study design, an individual patient has periods in which one receives the treatment and periods in which one does not. It must be possible to provide a meaningful placebo during the latter periods so both the patient and psychiatrist can be kept unaware of whether the patient is or is not receiving the treatment at any given time. This is obviously much easier to achieve with pharmacologic treatments than with psychotherapeutic ones. Also, the treatment needs to have an effect that has a relatively rapid onset as well as a relatively rapid loss of benefit when discontinued. This limits its applicability to most psychotherapeutic interventions, for example.

Achieving all these conditions, even where possible, can impose a significant burden on an individual clinician. Over the years a number of centres have been created to provide assistance to practitioners wishing to conduct N of 1 studies. Sadly, these centres have seldom survived for very long due to loss of funding and lack of sustained interest.[64] The reinvigorating of such programmes and the development of other strategies to allow meaningful information from the treatment of individual patients to be effectively collected and made available in a usable form is badly needed if there is to be any hope that the experiences of clinicians with individual patients can have a meaningful impact on the directions taken by the traditional research community.

We have argued that scientific method is not a unique form of thinking that can be defined in contrast to other effective human problem-solving strategies. Rather, the striking success of science must be found in the rigours and constraints that scientists bring to bear as they engage in solving the concrete problems and puzzles that they set for themselves. Prominent among these are the articulation of testable predictions about concrete situations and the engagement of the individual scientist with the efforts of others in their scientific community. Both apply fully to the work of the psychiatrist in constructing and testing POP models for understanding and treating individual patients. When done correctly and conscientiously, this work is a fully scientific endeavour.

Endnotes

1. Jaspers [1913] 1997
2. Palmer 1969; Phillips 1996
3. Makkreel 1992
4. Charon 2001
5. Anjum *et al.* 2020
6. Charon 2001
7. Godfrey-Smith 2021, pp. 4–5
8. Carnap 1995, pp. 3–18; Hempel 1966, pp. 47–69
9. Suppe 1974. pp. 3–118
10. The discussions that follow on the evolution of how scientific method has been understood relies heavily on work by the philosophers of science Nancy Cartwright, 1983,1999; Ian Hacking, 1983; Bas van Fraassen, 1980, 2008; Sandra Mitchell, 2003; Peter Godfrey-Smith, 2021 and especially Ronald Giere, 1988, 1999, 2006
11. Godfrey-Smith 2021, pp. 17–23
12. Smith 2008
13. Westfall 1971
14. Giere 1999, pp. 86–96
15. Austin 1970
16. Hume [1748] 1999
17. Russell [1912] 2013
18. Laplace [1814] 1951, p. 4
19. Cartwright 1999
20. Cartwright 1983
21. Cartwright 1999
22. Laudan 1983; Resnik 2000
23. Popper 1963

24. Quine 1996
25. Salmon 1989
26. Carnap 1995
27. Hempel 1966
28. Salmon 1989
29. Quine 1951
30. Reichenbach 1938
31. Schickore and Steinle 2006
32. Sharpin 1982; Pickering 1992; Longino 2020
33. Kuhn 1970
34. Giere 1999. The naturalism proposed here is a methodological naturalism that makes no claims about whether or not there exist forces or entities in the universe that in principle lie beyond the natural world and its explanatory principles. To *a priori* deny such possibilities is a metaphysical naturalism. Methodological naturalism simply asserts that the most productive way to explore anything is our world, including the workings of science itself, is an empirically based exploration of the phenomena of interest including employing scientific methods
35. Giere 1988
36. Giere 1988
37. Hacking 1983; Cartwright 1999; Woodward 2005
38. Giere 1988, p. 66
39. In physics conservative forces are those in which the work done by the force is independent of the path taken, or phrase it differently, if the work done over a closed path is zero. Examples include gravitational force, spring force, and electrostatic force. Examples of non-conservative forces are friction, air resistance, and tension in the material used. For different purposes, models of the same concrete piece of reality might be either conservative or non-conservative. Thus, a model for the motion of a pendulum that ignores air resistance and friction between the string and the pivot would be a conservative model, whereas a model that considered these forces would be non-conservative
40. Giere 1999
41. Giere 1988, p. 85
42. Giere 1988, p. 81
43. Giere 2006, p. 64
44. Mitchell 2003, p. 9
45. Giere 1988, p. 89
46. Giere 1988, pp. 89–90
47. Giere 1988
48. Cartwright 1983, 1999
49. Mitchell 2003
50. I thank John Sadler for drawing my attention to this example
51. Popper 1990
52. Cartwright 1999
53. Cartwright 1999, p. 77
54. Chase 2011
55. Cartwright 1999, p. 78
56. Cartwright 1999, p. 81

57. Woodward 2005
58. Kendler and Campbell 2009
59. Kind 2023
60. Plutynski 2024
61. Longino 2020
62. Berkenkotter 2008
63. Weisberg 2014
64. Mirza *et al.* 2017

Chapter 11
Conclusions and future directions

11.1 Reviewing the journey

It is time to review and summarize the journey we have made in this book: why we did it, the paths we have taken, the conclusions we have reached, and the significance of those conclusions for the practice of psychiatry. In Chapter 1, I made the case for an urgency to explicitly articulate and defend a methodology by which psychiatrists make treatment decisions with individual patients, as clinical decision-making in psychiatry and medicine in general has largely been a tacit activity. In recent decades, changes in the field of psychiatry have led to a devaluing of the clinical judgements of the practising psychiatrist about an individual patient. Increasingly the clinician is seen as the recipient of the results of research trials which are to guide clinical decision-making, with pride of place to the randomized controlled trial (RCT). This knowledge is expressed in terms of probabilistic outcomes for groups of patients. Individualized knowledge about a unique patient is devalued in contrast to information from these presumably more scientific sources. Only when such information is not yet available should the clinician be thrown upon intuitive judgement, the 'art of medicine'.

The pessimism resulting from this devaluation of clinical judgement is compounded by the disappointing and repeated failure to date of this same 'scientific psychiatry', which is supposed to guide us in our clinical decisions with patients, to live up to the rather grand expectations for categorically improving our understanding and treatment of the conditions that afflict our patients.

The evidence-based medicine (EBM) movement, originally limited to suggesting methods for the busy clinician to efficiently access and assess the large body of information reported in the literature, has subsequently been put forth as providing a systematic methodology for the clinician by applying the results of published research in making decisions about the treatment of the individual patient. We first asked if this could serve as the methodology we were seeking. In Chapter 2 we looked extensively at the procedures suggested by EBM to achieve this goal. I argued that the proposed methodology is

How Psychiatrists Make Decisions. Douglas W. Heinrichs, Oxford University Press. © Oxford University Press (2025).
DOI: 10.1093/9780198960720.003.0011

woefully inadequate to this task. It requires many arbitrary and unjustified assumptions and relies extensively on pseudo-quantification to make the procedure appear rigorous and meaningful. There are major problems with external validity. Many potentially biasing factors determine what gets subjected to controlled clinical trials and how those trials are structured. The psychiatrist is encouraged to dismiss much of the unique knowledge possessed about the patient. Ironically, there are no controlled studies to assess whether the application of EBM methodology itself actually improves patient outcomes. In short, EBM does not provide a blueprint for the methodology we need.

I posited however that there is a definable if currently largely implicit methodology already employed by experienced clinicians in making individualized treatment decisions, and that this method can be articulated and characterized. As a way of bringing forth this methodology, each of Chapters 3 through 6 looked in detail at an individual patient treated by me. The limitations of EBM were considered in each case. An effort was then made to make explicit the thinking process that led to treatment decisions, including revisions in that process when circumstances for the patients changed. In examining my thinking process in treating these cases, I in no way am suggesting that this process is unique to me. Quite the contrary, I think that most experienced psychiatrists would discover a similar way of reasoning if they attempted to make their own clinical thinking explicit. We noted that studies of expert reasoning in a range of fields demonstrate that it is possible to bring to conscious awareness the thinking process of an expert that was originally tacit and experienced as simply an intuition. I have attempted to engage in such a process in considering these cases.

An obvious limitation of this work to date is that my cases are all based on my experience of treating patients in a private practice setting. While I have no reason to think that a fundamentally different methodology of clinical reasoning operates in other settings, one important area for future work would be to apply the same strategy with patients being treated in other settings; such as acute and chronic inpatient units, forensic settings, public mental health clinics, consultation liaison psychiatry, and substance abuse programmes.

I then used the insights from the case studies to provide a more generalized characterization of pattern of propensities (POP) models organized around three proposed functions of those models: a descriptive function of how clinicians in fact reason in making individual decisions (Chapter 7), a pedagogical function of how the notion of POP models could enhance psychiatric training and the development of psychiatric expertise (Chapters 8 and 9), and an epistemological function of how casting clinical reasoning explicitly in terms of POP models can facilitate the validation of our assertions about a patient as credible knowledge and as instances of legitimate science (Chapter 10).

11.2 Major conclusions

These discussions cover a great deal of diverse terrain. As a summary, I have attempted to distil the major conclusions from this journey in the following 30 points:

1. A **POP model** is a mental construct that posits a number of **propensities and patterns of influence** between them that is intended to offer, looking backward, an **explanation** as to what brought about the patient's distress, and, looking forward, to suggest potential therapeutic **interventions** to relieve that distress in the future.
2. POP modelling is not offered as a new way of thinking, but as an articulation of the way good psychiatrists tacitly reason about patients. As such it is intended to make explicit the process of clinical decision-making. This is the **descriptive function** of POP models.
3. While model construction focuses specifically on the process of decision-making, it assumes and is built upon **other critical skills** possessed by the psychiatrist, including: familiarity with the knowledge base of the field, skill in eliciting relevant clinical information from interactions with the patient, and the capacity to establish a trusting alliance that allows the patient to share in an open way.
4. A model can be expressed in many forms such as a narrative or a diagram. The **narrative** has the advantage of making it easier to articulate complex and nuanced factors in detail. The **diagram** can facilitate a grasp of the overall pattern of the model more quickly and effectively, while sacrificing some detail and nuance. How the model is represented must always be distinguished from the model itself as a **mental construct.**
5. A POP model is the product of the disciplined and informed **imagination** of the clinician. The clinician draws upon a **store of exemplars** derived from relevant generalized knowledge obtained from the clinical and research literature and from the training received from supervisors and more senior clinicians. With growing experience, the psychiatrist develops one's own assembly of patterns based upon the patients one has treated that can function as exemplars as well.

6. POP models are always organized in terms of a specific **goal or goals** that the patient brings to treatment. Developing an explicit shared understanding of those goals with the patient is a critical element in the successful use of a model. Articulating a **shared model** with the patient requires that patients see the model as a credible explanation of their distress and a useful generator of therapeutic interventions. Using a model in this way encourages a true collaboration with the patient and guards against undue paternalism in the psychiatrist.
7. All models must be viewed as **provisiona**l and open to revision considering changes in the patient's goals or circumstances, unexpected outcomes of interventions, responses to other perturbations, or the acquisition of new history.
8. POP models are **pragmatic constructs**, judged by their effectiveness in bringing about desired results for the patient. There is no one objectively correct or 'true' model. Psychiatrists with different training, theoretical backgrounds and accumulated experience are likely to generate different models for the same patient that suggest different treatment interventions. This need not be problematic, as in most cases there are more than one set of interventions that would offer effective relief to the patient. It is critical, however, for the psychiatrist to generate **concrete and specific predictions** from the model whenever possible. The success or failure of those predictions ultimately determine the credibility of the model.
9. Even with a highly specific model about an individual patient, the model is an **abstraction** from the full complexity of the patient's life and circumstances. As such, it only approximately represents the patient's reality, and is never fully accurate. Whether it is similar enough to that reality depends upon whether it provides sufficient guidance for effective interventions to bring about the patient's goals. Models have been often likened to **maps**. Maps are abstract representations of various aspects of a complex geographic reality. Which elements they choose to represent and with what level of precision depends on the purpose for which the map is to be used.
10. In developing a model, it is helpful to look for **recurrent patterns** of distress or dysfunction that have characterized the patient over their lifetime, and the sorts of factors that have either brought about or ameliorated those patterns. In most cases the model is not just explaining a singular event, but a recurrent pattern in the patient's life.
11. POP models are described as collections of **propensities and patterns of interaction** between them, rather than in terms of **mechanisms**

and causes, to emphasize that outcomes are tremendously context dependent, as well as to emphasize the theoretical neutrality of models with regards to questions of determinism and existential freedom.

12. A model of a patient must be compatible with current generalized knowledge in the field. But the generalized knowledge does not directly translate into a recommended treatment for a specific patient. Rather **evidence** is used in a more fine-grained way to support the role of a specific propensity or path of influence between propensities that constitute a part of the model for that patient. Similarly, a patient is not simply an instance of a diagnosis. Rather a **diagnosis** functions as a placeholder to allow the positing of some components of the model so that evidence in the literature can be brought to bear even when the mechanism is currently poorly understood. This use of evidence contrasts sharply with the way evidence is applied in the approach recommended by EBM.
13. While models are the product of the psychiatrist's disciplined imagination, there are multiple important **constraints** on which models are acceptable. Hence from the range of possible models there is a **process of pruning** whereby many of those possible models are excluded**. Epistemological pruning** refers to the need for a model to be compatible with our best current knowledge in the field. **Pragmatic pruning** is needed when otherwise useful interventions are not available or cannot be effectively delivered in the patient's current circumstances. **Negotiated pruning** results from the need for the model to be acceptable to both doctor and patient as the basis of a shared understanding of the patient's distress and to be compatible with the patient's values and preferences. **Ethical pruning** refers to the need for the psychiatrist, in line with the mandate to 'first do no harm' to anticipate potential harmful outcomes from interventions suggested by the model resulting either from possible side effects or from consequences of the model being incorrect. Pruning requires the psychiatrist to consider a full range of 'what if' scenarios and counterfactuals, including the possibility that the current working model is wrong. The psychiatrist's **imaginative capacity** must be accompanied by the virtues of **clinical humility and honesty**.
14. In a field like psychiatry, models are quantitative only in the crude sense that one propensity is posited to either increase or decrease another propensity. Any attempt to be more precise about the magnitude of that effect will be extraordinarily approximate.

This reflects the current state of our knowledge, and honestly acknowledging this is much preferable to engaging in **pseudo-quantification** in an attempt to look more precise than we are. Furthermore, in assessing our confidence that a model or some component of that model is provisionally correct, we draw upon a broader and well-established **concept of probability** not limited to just what can be subjected to statistical and quantitative analysis.

15. The model for a specific patient can draw upon elements from various **levels of explanatory abstraction.** While at a theoretical level there is considerable debate as to whether some levels of explanation can be reduced to more basic ones, an individual model is free to combine elements from different levels based upon the pragmatic advantage of doing so in generating useful interventions. These elements must be combined into a unique model of the patient that generates very specific proposed interventions, and not simply a general admonition that we should consider a combination of biological, psychological, and social factors in the spirit of what has been termed the **biopsychosocial model**. What is needed is a targeted **pluralism** as opposed to a vacuous **eclecticism.** A **'manipulationist' or 'interventionist' theory of causal explanation** has been developed that is highly compatible with this approach, and is likely to be particularly fruitful in a field like psychiatry.
16. A potential value in constructing explicit POP models is that they can serve to clarify and at times resolve **differences between multiple clinicians** involved in the care of the same patient.
17. In the last half-century, a view of **scientific method** as a unique form of human thought designed to generate **laws of nature** and other uniform **generalizations** extending over some domain of study has been challenged as both conceptually inadequate and not an accurate reflection of how scientists reason. A more **naturalistic methodology** has looked at how scientists actually operate across disciplines. The resulting understanding sees scientific method not as a unique species of thought, but as a particularly rigorous application of sound human reasoning in general. The rigour derives in large part from theories being open to correction and modification based upon their success or failure in fulfilling **concrete and specific predictions** deriving from theories. The rigour also comes from the **public and communal nature of science**, in which the thinking of the scientist is open to critique and attempts to replicate by other scientists.

18. Studies of **general human reasoning** have evolved from seeing the roles of formal logic and probability as the gold standard, to a realization that many real-life problems are better solved with the intelligent application of **heuristics**. This is especially true when problems are complex, data is incomplete, considerable ambiguity exists, and there are practical limits on the time and energy available to solve the problem. In many situations looking for an adequate solution, a **satisficing strategy**, is preferable to a search for the best solution, and **optimizing strategy**. POP modelling is a satisfying strategy; EBM is an optimizing one.
19. Research focusing on the process by which experts in a range of fields develop their **expertise** suggests that they progressively move toward satisficing strategies and reliance on heuristics as they gain more experience. They ultimately come to rely on a **recognition-primed decision (RPD) strategy** in which the expert rapidly intuits from his store of experiences whether the current situation is a prototypical pattern or has anomalies. If it is prototypical, the expert virtually instantaneously generates a model of the relevant elements in the situation and how they can be transformed by other factors including interventions by the expert, an **action plan**. If anomalies exist, this step will take more time but remains relatively rapid. The expert then engages in a **mental simulation** in which those transformations are imagined. If the interventions play out mentally, the expert will initiate them. If they run into a problem, the model is revised and is again subjected to a mental simulation until a reasonable course of action is found. This process of recognition-primed decision-making is simply a rephrasing of the POP modelling process. In someone with a high level of experience and expertise, these processes, especially in prototypical cases, can happen so rapidly that the expert is not consciously aware of the cognitive processes involved but experiences them as **intuition.** This accounts for the tendency for the thinking process to remain tacit. But with appropriate reflection the expert can uncover the steps in their reasoning.
20. This research on the development of expertise has led to the articulation of a range of strategies for the training of novices in various fields to become experts. Conscious use of the methodology of POP modelling can facilitate a range of approaches that can be applied in psychiatric training to encourage the development of psychiatric experts. This is the **pedagogical function** of POP models. In contrast, EBM is an optimizing strategy seeking the 'best' treatment

in light of current knowledge. In addition to being unwieldy and often intractable, it risks encouraging the psychiatrist to continue to approach problems as a novice rather than evolve the strategies characteristic of an expert.

21. Some of the **pedagogical strategies** that could enhance the development of expertise in psychiatric training include having trainees explicitly articulate POP models for at least some of their patients in supervision, make specific predictions based on those models and carry out mental simulations. Particular attention should be placed on identifying **leverage points and choke points**. Optimal benefit from this requires that trainees have access to **meaningful feedback** whenever possible for the patients they treat. Trainees can be encouraged to conduct **'pre-mortem' exercises** to consider how a model may go wrong. Each step in the POP model can be assessed in terms of the **type and strength of the evidence** that supports it. These strategies constitute a set of preliminary suggestions that can form the basis of **a future pedagogical research programme in psychiatric training.**
22. A preliminary **taxonomy of the types of evidence** relevant to our confidence in any component of a POP model include data from controlled or large-scale studies, notions based on theories of underlying mechanisms, recurrent patterns that the clinician has observed in treating other patients that somewhat resemble the patient in question, communally shared experiences of other clinicians in the field and patterns familiar to them, and finally the use of concepts from folk psychology which overlaps with Jasper's concept of *verstehen*.
23. The **naturalistic studies of scientific method** suggest that working scientists across scientific disciplines predominantly approach their work from the vantage of **solving concrete problems or puzzles**. They do so by generating **models of a piece of concrete reality** they are attempting to understand or manipulate. These models are never literally true or a full reflection of the concrete reality they intend to model. The model is judged based upon it being sufficiently like the concrete reality in the relevant ways to meet the goals of the scientist in this specific instance. As experience with concrete problems in a particular area of interest accumulates, certain types of problems and successful solutions to those problems become apparent, and serve as **exemplars** for how one might solve future problems of a similar type. At times this leads to increasingly abstract models that suggest strategies for approaching problems in an ever-widening domain, **families of models**. In some scientific disciplines this can yield extremely

generalized understanding approaching something like laws of nature. But in many scientific disciplines this level of abstraction is never reached. But across all scientific disciplines, the more abstract models cannot be judged on their own merit, but only to the extent that they provide useful approaches for the construction of new models of concrete pieces of reality. These particular models are foundational in all scientific disciplines. In this view of the matter, science is fundamentally rooted in and justified in terms of constructing successful models that solve concrete problems.

24. Given this view of science deriving from a naturalistic philosophy of science, the psychiatrist engaging in constructing and assessing a POP model for an individual patient is engaged in **the paradigmatic activity of science**. The so-called art of medicine, properly understood, is really the science of medicine. This view of science is fully compatible with **hermeneutic strategies** for understanding and interpreting patients' experiences as well as with the use of **narrative** in bringing forth the uniqueness of each patient. The fruitless opposition of 'scientific' and 'nonscientific' medicine is rooted in the narrow concept of science espoused by the logical empiricists.
25. A further implication of the naturalistic view of science is that each scientific discipline must develop the specific methodologies and strategies that best suits its subject matter. POP modelling is an attempt to articulate the appropriate methodology for **psychiatry proper**, that is the effort of the psychiatrist to understand the distress or dysfunction of the unique individual and develop appropriate interventions to ameliorate the problem. It is useful to distinguish psychiatry proper from **other informing disciplines** that, while providing important information for the psychiatrist, have their own best methodologies. It is particularly important to distinguish psychiatry proper from **psychopathology** and **therapeutics**, which can be easily conflated with it but have their own methodologies.
26. In psychiatry proper, each patient is a **domain of study**, in essence the subject of their own small scientific discipline. The POP model for that patient is looking to explain certain recurrent patterns in that domain that constitute the distress or dysfunction for that patient. When one is developing and working with a POP model for Ms Jones, one is literally engaged in 'Jonesology'. Components of the model for Ms Jones may suggest useful approaches in constructing a model for Ms Smith, but the resulting model will be unique to Ms Smith, explaining recurrent patterns in the domain of 'Smithology'.

27. Developments that could add **rigour** and strengthen **the credibility and productivity of psychiatric science** include: formulating precise and testable **predictions** for the models we posit about a patient whenever possible; build mechanisms to allow prompt and specific **feedback** as to the outcome of interventions over time; supply increasing opportunities for **multiple clinicians** to examine and critique a proposed model for a given patient; and encouraging a **two-way exchange between clinicians and researchers** so that models of individual patients can enrich the thinking of researchers as well as researchers providing generalized data that may be of use to clinicians in generating future POP models.
28. In our culture, including medicine, being scientific is generally viewed as providing the pinnacle of credibility for a knowledge claim. The realization that the psychiatrist's reasoning about an individual patient is literally and fully a scientific activity not only enriches our understanding of clinical reasoning. It also is a compelling argument for restoring **the status of clinical judgements** about an individual patient as having considerable **epistemological credibility.**
29. The reflections that led to the articulation of **POP model construction** as the methodology of clinical reasoning and decision-making in psychiatry run parallel to two other research projects: 1. the work on characterizing both how expertise is developed and **the decision-making process of experts** across fields, and 2. the naturalistic approach to the philosophy of science leading to a **model-based view of scientific method.** The striking similarity in the conclusions reached by each of these pathways adds considerable credibility to the POP model approach.
30. The project of this book demonstrates the importance for psychiatrists to consciously reflect on the fundamental and often tacit concepts that underlie the way we understand what we do in day-to-day practice. Psychiatry deals with some of the most conceptually confusing and controversial questions about what it is to be a person with both physical and mental properties, to be both an agent and an object in the world. More than in most fields, we can easily be led astray by **conflicting or incoherent underlying assumptions.** The growing field of **the philosophy of psychiatry**[1] can provide considerable help in navigating this terrain and in developing a basic **conceptual competence** for the clinician. Some exposure to this work as part of psychiatric training would facilitate the psychiatrist's ongoing reflections throughout

their career on these critical issues. **Model curriculums** have already been put forth to address this need.[2]

11.3 Some future directions

This book is my preliminary effort to articulate the methodology of clinical reasoning and decision-making employed by psychiatrists in working with individual patients. The result is the notion of a POP model. I believe that such a model both fairly characterizes the thinking of experienced clinicians and has useful implications for understanding what we do, for enhancing the training of future psychiatrists and the ongoing growth of more experienced psychiatrists, and for justifying our conclusions about individual patients as valid knowledge. As with any preliminary effort, it has many limitations. I hope it will generate enough interest to encourage future work to address those limitations. Before closing, I want to indicate a few of the most obvious areas that need to be addressed.

A POP model approach needs to be studied with patients in settings other than private practice, and with psychiatrists with training backgrounds and theoretical perspectives much different from my own. How different are the models that emerge in these contexts from what I have presented here? Do new features or challenges for model construction emerge in these settings? Is my conceptual framework for understanding these models sufficiently broad to encompass them?

Much more work needs to be done on how to use POP models to enhance psychiatric training. How can supervisors use this framework most effectively? How can effective feedback as to the accuracy of predictions be best provided given the many practical constraints of psychiatric training? What kinds of clinical simulations would be most helpful in teaching trainees how to develop a model? Further work is needed to develop a comprehensive taxonomy of the kinds of evidence that can be used to assess the strength of each proposed component of a POP model. How can we best engage a patient in a truly collaborative process of formulating a shared POP model? What are the most effective ways to teach a trainee how to question a model and modify or change it when appropriate? What is the best way to integrate the use of diagnoses in our understanding of POP models? How can we find room for some basic teaching of conceptual competence (philosophy of psychiatry) among the extensive demands and time constraints placed on psychiatric trainees? How do we encourage meaningful two-way exchanges between clinicians and professional researchers in psychiatry?

At various points in my discussion, some of the ethical implications of employing POP models were discussed. Much more should be explored about these. Viewing model construction as a scientific activity makes the treatment of each patient based on predictions from the model its own scientific experiment. Cognitively, I think this is a very useful way to conceptualize the matter. However, it is extremely important to remember that this is not an experiment run in a test tube, but involves decisions that profoundly and potentially affect, for better or worse, the lives of our patients. The onus is on the psychiatrist to consider potential adverse effects intrinsic to the treatments we propose as well as unintended consequences when our model is wrong. How should we best use principles of informed consent to bring the patient into the decision-making process as fully as possible, without using it to deflect from the fact that as treating clinicians the anticipation of and prompt response to any harm is our responsibility? More work on characterizing the process of ethical pruning in model construction is needed.

How does the use of a POP model framework impact the development of clinical virtues in the psychiatrist?[3] My own belief is that there is a fundamental difference in this regard between this approach and that based on EBM, in that the latter starts from what is assumed to be the best current knowledge relevant to treating the patient, whereas the POP approach posits that the optimal treatment cannot be defined until the patient's goals are clearly articulated and mutually agreed upon between patient and doctor. I believe that the EBM approach encourages a 'doctor knows best' stance and then tries to tack on, rather awkwardly, a consideration of the patient's values and preferences. An inevitable bias toward paternalism results. There is a strong tendency for the clinician to see the task as persuading the patient that employing the predefined best treatment is desirable. In contrast, employing a POP model involves open collaboration and transparency from the start. We begin from patients' frames of reference as reflected in their goals. Speculation as to the best treatment cannot begin prior to a consideration of these goals. Hence it is my conviction that using POP modelling encourages the clinical virtues of humility, transparency, and collaboration. Having said all of this, however, much further empirical work needs to be done to determine whether these assertions prove true.

I have drawn considerable parallels between POP models in psychiatry and the way models are used in other scientific disciplines. Considerably more work needs to be done in refining our understanding of the extent to which POP models fit the general pattern of scientific models as well as identifying features that may be unique to psychiatry. One obvious area needing further exploration follows from the fact that with a model-based view of science, scientific method is no longer viewed as a unique species of human thought but as a particularly rigorous and disciplined form of normal human reasoning. The credibility of work in any field of science hinges on the disciplined patterns of thought of the scientist as well as the institutional structures that promote that rigour. The thought patterns of the individual scientist involve constructing models that lend themselves to precise and discriminating predictions for which clear feedback can be obtained in a reasonable timeframe. Models must also be compatible with the body of pre-existing knowledge in the field. Institutional structures should maximize the opportunity for the models generated by the individual scientist wrestling with a specific problem to be exposed to the scrutiny and critique of other members of that scientific community. Much work needs to be done in psychiatry to strengthen both these pillars of rigour.

11.4 Some final thoughts

In the introduction I stated that this book was addressed to two groups of readers. The primary audience is composed of practising psychiatrists, psychiatric residents, psychiatric educators, and other mental health clinicians. Preliminary though it is, I hope this effort encourages a better understanding of how we reason about patients and the value of making that understanding explicit, as well as an appreciation for the need to examine the tacit conceptual foundations of how we think about our discipline. I hope it leads to further advances in training and enhanced expertise. A second audience consists of philosophers of psychiatry. I hope this work provides useful reflections on how psychiatrists wrestle with the day-to-day challenges of clinical care, and thereby makes the conceptual work of the philosopher more grounded in clinical reality. Finally, I hope that this project is a step toward reversing the devaluation of clinical judgement in contemporary psychiatry, justifying the results of that judgement as valid if fallible scientific knowledge, and thereby encouraging practising psychiatrists to have more confidence, pride, and satisfaction in what we do in treating patients.

Endnotes

1. There is a rapidly expanding literature in the philosophy of psychiatry in the form of books and journals aimed at psychiatrists and philosophers. An especially comprehensive textbook is the *Oxford Textbook of Philosophy and Psychiatry* 2006 (Fulford, Thorton,

and Graham). A stimulating interdisciplinary organization composed largely of psychiatrists and philosophers and very welcoming to psychiatrists and trainees new to this field is the Association for the Advancement of Philosophy and Psychiatry. https://aapp.press.jhu.edu/

2. Aftab and Waterman 2021
3. Radden and Sadler 2010

References

Aftab, A. and G. S. Waterman. 2021. Conceptual competence in psychiatry: recommendations for education and training. *Academic Psychiatry* 45:203–209.

Agency for Healthcare Research and Quality. 2010. *Comparative Effectiveness of Second-Generation Antidepressants in the Pharmacologic Treatment of Adult Depression*. Washington: Government Printing Office.

American Psychiatric Association. 1980. *Diagnostic and Statistical Manual of Mental Disorders third edition (DSM-3)*. Washington: American Psychiatric Publishing.

American Psychiatric Association. 2013. *Diagnostic and Statistical Manual of Mental Disorders fifth edition (DSM-5)*. Washington: American Psychiatric Publishing.

Anjum, R. L., S. Copeland, and E. Rocca (Eds.) 2020. *Rethinking Causality, Complexity and Evidence for the Unique Patient: A CauseHealth Resource for Healthcare Professionals and the Clinical Encounter*. Cham: Springer.

APA Presidential Task Force on Evidence-Based Practice. 2006. Evidence-based practice in psychology. *The American Psychologist* 61(4):271–285.

Aristotle. 1999. *Nicomachean Ethics*, second edition. Irwin, T. (trans). Indianapolis: Hackett.

Austin, W. H. 1970. Isaac Newton on science and religion. *Journal of the History of Ideas* 31(4):521–542.

Begg, C. B. 1994. Publication bias. In Cooper, H. and I. V. Hedges (eds). *The Handbook of Research Synthesis*. New York: Russell Sage Foundation, pp. 400–408.

Benazzi, F. 2007. Bipolar II disorder. *CNS Drugs* 21:727–740.

Beck, A. T. *et al.* 1961. An inventory for measuring depression. *Archives General Psychiatry* 4:561.

Berk, M. and S. Dodd. 2005. Bipolar II disorder: a review. *Bipolar Disorders* 7(1):11–21.

Berkenkotter, C. 2008. *Patient Tales: Case Histories and the Uses of Narrative in Psychiatry*. Columbia: University of South Carolina.

Bethany, M., *et al.* 2010. Treatment preference, engagement, and clinical improvement in pharmacotherapy versus psychotherapy for depression. *Behaviour Research and Therapy* 48(8):799–804.

Bhandari, M., *et al.* 2004. Association between industry funding and statistically significant pro-industry findings in medical and surgical randomized trials. *Canadian Medical Association Journal* 170(4):477–480.

Bluhm, R. and K. Borgerson. 2011. Evidence-based medicine. In Gifford, F. (Ed.), *Handbook of the Philosophy of Science. Volume 16: Philosophy of Medicine*. Elsevier BV, pp. 203–238.

Burks, A. W. 1946. Peirce's theory of abduction. *Philosophy of Science* 13(4):301–306.

Carnap, R. 1995. *An Introduction to the Philosophy of Science*. New York: Dover.

Cartwright, N. 1983. *How the Laws of Physics Lie*. New York: Oxford.

Cartwright, N. 1999. *The Dappled World: A Study of the Boundaries of Science*. Cambridge: Cambridge University Press.

Casey, B. J., *et al.* 2013. DSM-5 and RDoC: progress in psychiatry research? *Nature Reviews Neuroscience* 14(11):810–814.

Charon, R. 2001. Narrative medicine: a model for empathy, reflection, profession, and trust. *JAMA* 286(15):1897–1902.

Chase, M. 2011. Teleology and final causation in Aristotle and in contemporary science. *Dialogue*, 50(3):511–536.

Churchland, P. S. 1982. Mind-brain reduction: new light from the philosophy of science. *Neuroscience* 7(5):1041–1047.

Cooper, R. 2004. What is wrong with the DSM? *History of Psychiatry* 15(1):5–25.

Cosmides, L. 1989. The logic of social exchange: has natural selection shaped how humans reason? studies with the Watson selection task. *Cognition* 31(3):187–276.

Cuthbert, B. N. and T. R. Insel. 2013. Toward the future of psychiatric diagnosis: the seven pillars of RDoC. *BMC Medicine 11*(1):1–8.

Davidson, J. T. 2007. A history of the concept of atypical depression. *J Clin Psychiatry* 68 (supp 3):10–15.

Debnath, L. and K. Basu. 2015. A short history of probability theory and its applications. *International Journal of Mathematical Education in Science and Technology* 46(1):13–39.

Descartes, R. [1641]. 1996. Meditation VI. In *Meditations on the First Philosophy*, J. Cottingham (trans.). Cambridge: Cambridge University Press, pp. 57–72.

Dunlop, B., M. Thase, C. C. Wun, *et al.* 2012. A meta-analysis of factors impacting detection of antidepressant efficacy in clinical trials: the importance of academic sites. *Neuropsychopharmacol*ogy 37:2830–2836.

Dutton, D. G. and A. P. Aron. 1974. Some evidence for heightened sexual attraction under conditions of high anxiety. *Journal of Personality and Social Psychology* 30(4):510–517.

Engel G. L. 1977. The need for a new medical model: a challenge for biomedicine. *Science.* 196:129–136.

Engel G. L. 1980. The clinical application of the biopsychosocial model. *Am J Psych* 137: 535–544.

Fort, D., T. Herr, P. Shaw, K. Gutzman, and J. Starren. 2017. Mapping the evolving definitions of translational research. *Journal of Clinical and Translational Science* 1(1):60–66.

Frances, A. 2014. RDoC is necessary, but very oversold. *World Psychiatry* 13(1):47.

Fulford K. W. M., T. Thorton, and G. Graham. 2006. Oxford Textbook of Philosophy and Psychiatry. Oxford

Gaynes, B. N. 2009. Identifying difficult-to-treat depression: differential diagnosis, subtypes, and co-morbidities. *J Clin Psychiatry* 70 (Suppl 6):10–15.

Ghaemi, S. N. 2003. *The Concepts of Psychiatry: A Pluralistic Approach to the Mind and Mental Illness.* Baltimore: Johns Hopkins.

Giere, R. N. 1988. *Explaining Science: A Cognitive Approach.* Chicago: University of Chicago.

Giere, R. N. 1999. *Science without Laws.* Chicago: University of Chicago.

Giere, R. N. 2006. *Scientific Perspectivism.* Chicago: University of Chicago.

Gigerenzer, G. 2008. *Rationality for Mortals: How People Cope with Uncertainty.* New York: Oxford.

Gluud L. L. 2006. Bias in clinical intervention research. *American Journal of Epidemiology,* 163(6):493–501.

Godfrey-Smith, P. 2021 *Theory and Reality.* Chicago: University of Chicago.

Gold, I. 2009. Reduction in psychiatry. *Canadian Journal of Psychiatry* 54(8):506–512.

Goldberg, J. F. and S. M. Stahl, 2024. *Clinical Reasoning and Decision-making in Psychiatry.* Cambridge: Cambridge University Press.

Goldstein, D. G. and G. Gigerenzer 1999. The recognition heuristic: how ignorance makes us smart, In Gigerenzer G., *et al.* (eds) *Simple Heuristics That Make Us Smart.* New York: Oxford University Press, pp. 37–58.

Goze, T., M. Moskalewitz, M., A. Schwartz, *et al.* 2019. Reassessing 'praecox feeling' in diagnostic decision making in schizophrenia: a critical review. *Schizophrenia Bulletin* 45(5):966–970.

Gupta, M. 2014. *Is Evidenced-based Psychiatry Ethical?* Oxford: Oxford University.

Guyatt, G., D. Rennie, M. Meade, and D. Cook (Eds). 2001. *Users' Guide to the Medical Literature: A Manual for Evidence Based Clinical Practice.* Chicago: AMA Press.

Guze, S. 1989. Biological psychiatry- is there any other kind? *Psychological Medicine* 19(2):315–323.

Hacking, I. 1983. *Representing and Intervening: Introductory Topics in the Philosophy of Natural Science*. Cambridge: Cambridge University.

Hacking, I. 2006. *The Emergence of Probability: A Philosophical Study of Early Ideas about Probability, Induction and Statistical Inference*, second edition. Cambridge: Cambridge University.

Hamilton, M. E. 1960. A rating scale for depression. *Journal of Neurology, Neurosurgery, and Psychiatry* 23:56–62.

Harrow, M. T., H. Jobe, and L. Tong. 2022. Twenty-year effects of antipsychotics in schizophrenia and affective psychotic disorders. *Psychological Medicine* 52(13):2681–2691.

Healy, D. 1997. *The Anti-depressant Era*. Cambridge: Harvard.

Heinrichs, D. W. 2015. Model-based science and the ethics of ongoing treatment negotiations. In J. Z., Sadler, C. W. van Staden, and K.W.M. Fulford (Eds.), *The Oxford Handbook of Psychiatric Ethics*. Oxford: Oxford University, pp. 1143–1159.

Hempel, C. 1965. *Aspects of Scientific Explanation*. New York: The Free Press.

Hempel, C. 1966. *Philosophy of Natural Science*. Upper Saddle River: Prentice Hall.

Hoffrage, U., S. Lindsey, R. Hertwig, and G. Gigerenzer. 2000. Communicating statistical information. *Science* 290(5500):2261–2262.

Hume, D. [1748] 1999. *An Enquiry Concerning Human Understanding*. Oxford: Oxford University Press.

Hutto, D. and I. Ravenscroft 2021. Folk psychology as a theory. *The Stanford Encyclopedia of Philosophy* (Fall 2021 Edition), E. N. Zalta (Ed.). https://plato.stanford.edu/archives/fall2021/entries/folkpsych-theory/.

Ierodiakonou, C. 2014. Medicine as a model for Aristotle's ethics and his person-centered approach, *International Journal of Person Centered Medicine* 4(1):31–34.

Jaeger, W. 1957. Aristotle's use of medicine as model of method in his ethics. *Journal of Hellenic Studies* 77:54–61.

Jaspers, K. [1913] 1997. *General Psychopathology*. (trans. Hoenig J. and M. W. Hamilton) Baltimore: Johns Hopkins.

Johns Hopkins School of Public Health. 2015. Industry financed clinical trials on the rise as number of NIH funded trials falls. https://www.jhsph.edu/news/news-releases.

Kahneman, D. 2011. *Thinking, Fast and Slow*. New York: Farrar, Straus, and Giroux.

Karlsson, H. and M. Kamppinen. 1995. Biological psychiatry and reductionism: empirical findings and philosophy. *British Journal of Psychiatry* 167(4):434–438.

Kendler, K. 2005. Toward a philosophical structure for psychiatry. *Am J Psychiatry* 162: 433–440.

Kendler, K. and J. Campbell. 2009. Interventionist causal models in psychiatry: repositioning the mind-body problem. *Psychological Medicine* 39:881–887.

Kendler, K. S. and K. F. Schaffner. 2011. The dopamine hypothesis of schizophrenia: an historical and philosophical analysis. *Philosophy, Psychiatry, & Psychology* 18(1):41–63.

Khan, A. and W. A. Brown. 2001. The placebo enigma in antidepressant clinical trials. *J Clin Psychopharmacology* 21:123–125.

Kind, A. 2023. How does the psychiatrist know? *Philosophy and the Mind Sciences* 4(6):1–38.

Kinghorn, W. 2020. Crisis in psychiatric diagnosis? Epistemological humility in the DSM era. *The Journal of Medicine and Philosophy: A Forum for Bioethics and Philosophy of Medicine* 45(6):581–597.

Kirk, S. A., D. Cohen, and T. Gomory. 2015. DSM-5: The delayed demise of descriptive diagnosis. In Demazeux S. and P. Singy (Eds.). *The DSM-5 in Perspective: Philosophical Reflections on the Psychiatric Babel*. Dordrecht: Springer, pp. 63–81

Klein, G. 1998. *Sources of Power: How People Make Decisions*. Cambridge: MIT.

Kuhn, T. 1970. *The Structure of Scientific Revolutions*, second edition. Chicago: University of Chicago.

Kwag, K. H., M. Ganzales-Lorenzo, R. Banzi R, *et al.* 2016. Providing doctors with high quality information: an updated evaluation of web-based point-of-care information summaries. *J Med Internet Res* 18(1):e15.

Lakatos, I. 1978. *Philosophical Papers: Volume 1*. Cambridge: Cambridge University Press.

Laney, M. and J. Sadler. 2021. Psychiatric diagnosis. In Bloch, S. and S. A. Green (Eds.). *Psychiatric Ethics*. Oxford: Oxford University Press. pp. 97–120.

Laplace, P S. [1814] 1951. *A Philosophical Essay on Probabilities* (Trans. Truscott F.W. and F. L. Emory). New York: Dover.

Larkin, J., J. McDermott, D. P. Simon, and H. A. Simon. 1980. Expert and novice performance in solving physics problems. *Science* 208(4450):1335–1342.

Laudan, L.1983. The Demise of the demarcation problem. In Cohen, R.S. and L. Laudan (Eds.). *Physics, Philosophy and Psychoanalysis. Boston Studies in the Philosophy of Science.* vol 76. Dordrecht: Springer, pp. 111–128.

Légaré, F., R. Adekpedjou, D. Stacey, S. Turcotte, J., *et al.* 2018. Interventions for increasing the use of shared decision making by healthcare professionals. *Cochrane Database of Systematic Reviews* 7:CD006732.

Liebowitz, M. R. and D. F. Klein. 1979. Hysteroid dysphoria. *Psychiatric Clinics* 2(3):555–575.

Lilienfeld, S. O. 2014. The Research domain criteria (RDoC): an analysis of methodological and conceptual challenges. *Behaviour Research and Therapy* 62:129–139.

Lillie, E. O., B. Patay, J. Diamant, B. Issell, *et al.* 2011. The n-of-1 clinical trial: the ultimate strategy for individualizing medicine? *Personalized Medicine* 8(2):161–173.

Longino, H. E. 2020. *Science as Social Knowledge: Values and Objectivity in Scientific Inquiry*. Princeton: Princeton University Press.

Luhrmann, T. 2001. *Of Two Minds: An Anthropologist Looks at American Psychiatry*. New York: Vintage.

Makkreel, R. A. 1992. *Dilthey: Philosopher of the Human Studies*. Princeton: Princeton University.

Mayes, R. and A. V. Horwitz. 2005. DSM-III and the revolution in the classification of mental illness. *Journal of the History of the Behavioral Sciences* 41(3):249–267.

Mercer, H. and D. Sperber. 2017. *The Enigma of Reason*, Cambridge: Harvard University Press.

Miczek, K. A. and T. W. Robbins. 2019. Psychopharmacology in its 60th year. *Psychopharmacology* 236:3383–3384.

Midgley, M. 2005. *The Essential Mary Midgely*. (Midgely D. Ed.). New York: Routledge.

Miller, A. and R. T. Ward. 1981. *Prisoners of Childhood: The Drama of the Gifted Child and the Search for the True Self*. New York: Basic Books.

Mirza R. D., S. Punja, and S. Vohra. 2017. The history and development of N-of1 trials. *Journal of the Royal Society of Medicine* 110(8):330–340.

Mitchell, S. D. 2003. *Biological Complexity and Integrative Pluralism*. Cambridge: Cambridge University.

Montori, V. M. and G. H. Guyatt. 2008. Progress in evidence based medicine. *JAMA* 300: 1814–1816.

Morrissey, J. P. and H. H. Goldman. 1984. Cycles of reform in the care of the chronically mentally ill. *Psychiatric Services* 35(8):785–793.

Munson, R. 1981. Why medicine cannot be a science. *The Journal of Medicine and Philosophy* 6(2):183–208.

Murphy, D. 2008. Levels of explanation in psychiatry. In Kendler, K. and J. Parnas (Eds.). *Philosophical Issues in Psychiatry: Explanation, Phenomenology and Nosology*. Baltimore: Johns Hopkins University, pp. 99–125.

Nutt, D. J. 2008. Relationship of neurotransmitters to the symptoms of major depressive disorder. *Journal of Clinical Psychiatry* 69(suppl E1):4–7.

Owens, D. 1989. Levels of explanation. *Mind* 98(389):59–79.

Pae, C. U., H. Tharwani, D. M. Marks, P. S. Masand, and A. Patkar. 2009. Atypical depression. *CNS Drugs* 23:1023–1037.

Palmer, R. E. 1969. *Hermeneutics: Interpretation Theory in Schleiermacher, Dilthey, Heidegger, and Gadamer*. Evanston: Northwestern University Press.

Phelps, J., J. Angst, J. Katzow, and J. Sadler. 2008. Validity and utility of bipolar spectrum models. *Bipolar Disorders* 10(1p2):179–193.

Phillips, J. 1996. Key concepts: hermeneutics. *Philosophy, Psychiatry, & Psychology* 3(1):61–69.

Phillips, J. and J. Z. Sadler. 2021. Evidence, science, and ethics in talk-based healing practices. In Trachsel M., *et al.* (Eds.). *The Oxford Handbook of Psychotherapy Ethics*. Oxford: Oxford University, pp. 288–311.

Pickering A. (Ed.). 1992. *Science as Practice and Culture*. Chicago: University of Chicago Press.

Pillinger, T. 2021. Treatment resistance in psychiatry: state of the art and new directions. *Molecular Psychiatry* 19:5.

Plutynski, A. 2024. How is a therapist like a modeler? *Philosophy, Psychiatry, & Psychology* 31(2):151–161.

Poland, J., and B. von Eckardt. 2013. Mapping the domain of mental illness. In Fulford *et al.* (Eds.). *The Oxford Handbook of Philosophy and Psychiatry*. Oxford: Oxford University Press, pp. 735–752.

Polanyi, M. 2009. *The Tacit Dimension*. Chicago: University of Chicago.

Popper, K. R. 1963. Science as falsification. *Conjectures and Refutations*. London: Routledge and Keagan Paul, pp. 33–39.

Popper, K. 1990. *A World of Propensities*. New York: Thoemmes.

Press, G. 1999. Origins of western philosophic thinking: introduction. In Popkin R. H. (Ed.). *Columbia History of Western Philosophy*. New York: MJF Books, pp. 1–5.

Quine, W. V. O. 1951. Two dogmas of empiricism. *Philosophical Review* 60:20–43.

Quine, W. V. O. 1996. Progress on two fronts. *Journal of Philosophy* 93:159–163.

Racagni, G. and M. Popoli. 2008. Cellular and molecular mechanisms in the long-term action of antidepressants. *Dialogues in Clinical Neuroscience* 10(4):385–400.

Radden, J. and J. Sadler. 2010. The Virtuous Psychiatrist: Character Ethics in Psychiatric Practice. Oxford University Press.

Reichenbach, H. 1938. *Experience and Prediction: An Analysis of the Foundations and the Structure of Knowledge*. Chicago: University of Chicago.

Resnik, D. 2000. A pragmatic approach to the demarcation problem. *Studies in History and Philosophy of Science* 31(2):249–267.

Rieskamp, J. and P. E. Otto. 2006. SSL: a theory of how people learn to select strategies. *Journal of Experimental Psychology: General* 135:207–236.

Ross, C. K., C. A. Steward, and J. M. Sinacore. 1993. The importance of patient preferences in the measurement of health care satisfaction. *Med Care* 31:1138–1149.

Rubio, D. M., E. E. Schoenbaum, L. S. Lee, D. E. Schteingart, *et al.* 2010. Defining translational research: implications for training. *Academic Medicine: Journal of the Association of American Medical Colleges* 85(3):470–475.

Russell, B. [1912] 2013. *The Problems of Philosophy*. New York: Simon and Brown.

Rutherford, B. R., E. Pott, J. M. Tandler, *et al.* 2014. Placebo response in antipsychotic clinical trials: a meta-analysis. *JAMA Psychiatry* 71(12):1409–1421.

Sackett, D. L., D. E. Strauss, W. S. Richardson, W. Rosenberg, and R. B. Haynes. 2000. *Evidence-based Medicine: How to Practice and Teach EBM*, second edition. New York: Churchill Livingstone.

Sadler, J. Z. 2005. *Values and Psychiatric Diagnosis.* Oxford: Oxford University.
Sadler, J. Z. 2013. Considering the economy of DSM alternatives. In Paris, J. and J. Phillips (Eds.). *Making the DSM-5.* Springer, New York. pp. 21–38.
Sadler, J. Z. 2021. Ethics considerations in selecting psychotherapy modalities and formats. In Trachsel, M., *et al.* (Eds.). *The Oxford Handbook of Psychotherapy Ethics.* Oxford: Oxford University, pp. 363–380.
Sadler, J. Z. and Y. F. Hulgus. 1992. Clinical problem solving and the biopsychosocial model. *American Journal of Psychiatry* 149(10):1315–1323.
Salmon, W. C. 1989. *Four Decades of Scientific Explanation.* Pittsburgh: University of Pittsburgh Press.
Sansone, R. A. and L. A. Sansone. 2010. SSRI-induced indifference. *Psychiatry* 7(10):14–18.
Schaffner, K. F. 1993. *Discovery and Explanation in Biology and Medicine.* Chicago: University of Chicago.
Schaffner, K. F. 2016. *Behaving: What's Genetic, What's Not, and Why Should We Care?* New York: Oxford University Press.
Schickore, J. and F. Steinle. 2006. Introduction: revisiting the context distinction. In Schickore, J. and F. Steinle (Eds.). *Revisiting Discovery and Justification: Historical and Philosophical Perspectives on the Context Distinction.* Dordrecht: Springer, pp. 7–19.
Scholl, I., A. LaRussa, P. Hahlweg, S. Kobrin, and G. Elwyn. 2018. Organizational-and system-level characteristics that influence implementation of shared decision-making and strategies to address them—a scoping review. *Implementation Science* 13(1):1–22.
Scull, A. 2022. *Desperate Remedies: Psychiatry and the Mysteries of Mental Illness.* Cambridge: Belknap Press.
Sharpin, S. 1982. History of science and its sociological reconstructions. *History of Science* 20:157–211.
Simon, H. 1955. A behavioral model of rational choice. *Quarterly Journal of Economics* 69: 99–118.
Singh, T. and K. Williams. 2006. Atypical depression. *Psychiatry* 3(4):33–39.
Slade, M. 2017. Implementing shared decision making in routine mental health care. *World Psychiatry* 16(2):146–153.
Smith, G. 2008. Isaac Newton. In Zalta, E. N. (Ed.). *The Stanford Encyclopedia of Philosophy (Fall 2008 Edition).* https://plato.stanford.edu/archives/fall2008/entries/newton/.
Sonawalla, S. B. and J. F. Rosenbaum 2002. Placebo response in depression. *Dialogues in Clinical Neuroscience* 4(1):105–113.
Spitzer, R. L. and J. B. Williams. 1982. Hysteroid dysphoria: an unsuccessful attempt to demonstrate its syndromal validity. *The American Journal of Psychiatry* 139(10):1286–1291.
Stomski, N. J. and P. Morrison. 2017. Participation in mental healthcare: a qualitative meta-synthesis. *International Journal of Mental Health Systems* 11(1):1–11.
Straus, S. E., P. Glasziou, W. S. Richardson, and R.B. Haynes. 2019. *Evidenced-based Medicine: How to Practice and Teach EBM,* fifth edition. Edinburgh: Elsevier.
Sullivan, G. M. and R. Feinn. 2012. Using effect size—or why the p value is not enough. *J Grad Med Educ* 4(3):279–282.
Suppe, F. 1974. *The Structure of Scientific Theories.* Urbana: University of Illinois.
Suppe, F. 1989. *The Semantic Conception of Theories and Scientific Realism.* Urbana: University of Illinois.
Swartz, M. S., T. S. Stroup, J. P. McEvoy, *et al.* 2008. What CATIE found: results from the schizophrenia trial. *Psychiatric Services* 59(5):500–506.
Thornton, S. 2021. "Karl Popper", *The Stanford Encyclopedia of Philosophy* (Winter 2023 Edition), Edward N. Zalta and Uri Nodelman (eds.) https://plato.stanford.edu/archives/win2023/entries/popper/.

Tong, E. K., T. P. Carmody, and J. A. Simon. 2006. Bupropion for smoking cessation. *Comprehensive Therapy* 32:26–33.
van Fraassen, B. C. 1980. *The Scientific Image*. Oxford: Clarendon.
van Fraassen, B. C. 2008. *Scientific Representation*. Oxford: Clarendon.
von Neumann, J. and O. Morganstern 1944. *Theory of Games and Economic Behavior*. Princeton: Princeton University Press.
Wallace E., J. Radden, and J. Z. Sadler. 1997. The philosophy of psychiatry: who needs it? *The Journal of Nervous and Mental Disease* 185(2):67–73.
Warden, D., A. J. Rush, and M. H. Trivedi, *et al.* 2007. The STAR*D project results: a comprehensive review of findings. *Current Psychiatry Reports* 9:449–459.
Weisberg, H. I. 2014. *Willful Ignorance: The Mismeasure of Uncertainty*. Hoboken: Wiley.
Westfall, R. S. 1971. *Force in Newton's Physics*. New York: American Elsevier.
Woodward J. 2005. *Making Things Happen: A Theory of Causal Explanation*. Oxford: Oxford University Press.

Index

For the benefit of digital users, indexed terms that span two pages (e.g., 52–53) may, on occasion, appear on only one of those pages.

Note: Figures are indicated by an italic *f* following the page number.